Designing for Human Intelligence in an Artificial Intelligence World

Understanding Human Cognition to Design for Humans

Jerome L. Rekart
Rebecca Baker

Apress®

Designing for Human Intelligence in an Artificial Intelligence World:
Understanding Human Cognition to Design for Humans

Jerome L. Rekart
Windham, NH, USA

Rebecca Baker
McKinney, TX, USA

ISBN-13 (pbk): 979-8-8688-1417-4
https://doi.org/10.1007/979-8-8688-1418-1

ISBN-13 (electronic): 979-8-8688-1418-1

Copyright © 2025 by Jerome L. Rekart and Rebecca Baker

This work is subject to copyright. All rights are reserved by the Publisher, whether the whole or part of the material is concerned, specifically the rights of translation, reprinting, reuse of illustrations, recitation, broadcasting, reproduction on microfilms or in any other physical way, and transmission or information storage and retrieval, electronic adaptation, computer software, or by similar or dissimilar methodology now known or hereafter developed.

Trademarked names, logos, and images may appear in this book. Rather than use a trademark symbol with every occurrence of a trademarked name, logo, or image we use the names, logos, and images only in an editorial fashion and to the benefit of the trademark owner, with no intention of infringement of the trademark.

The use in this publication of trade names, trademarks, service marks, and similar terms, even if they are not identified as such, is not to be taken as an expression of opinion as to whether or not they are subject to proprietary rights.

While the advice and information in this book are believed to be true and accurate at the date of publication, neither the authors nor the editors nor the publisher can accept any legal responsibility for any errors or omissions that may be made. The publisher makes no warranty, express or implied, with respect to the material contained herein.

Managing Director, Apress Media LLC: Welmoed Spahr
Acquisitions Editor: Shiva Ramachandran
Development Editor: James Markham
Project Manager: Jessica Vakili

Distributed to the book trade worldwide by Springer Science+Business Media New York, 1 New York Plaza, New York, NY 10004. Phone 1-800-SPRINGER, fax (201) 348-4505, e-mail orders-ny@springer-sbm.com, or visit www.springeronline.com. Apress Media, LLC is a Delaware LLC and the sole member (owner) is Springer Science + Business Media Finance Inc (SSBM Finance Inc). SSBM Finance Inc is a **Delaware** corporation.

For information on translations, please e-mail booktranslations@springernature.com; for reprint, paperback, or audio rights, please e-mail bookpermissions@springernature.com.

Apress titles may be purchased in bulk for academic, corporate, or promotional use. eBook versions and licenses are also available for most titles. For more information, reference our Print and eBook Bulk Sales web page at http://www.apress.com/bulk-sales.

If disposing of this product, please recycle the paper

JLR: "To Kathy, thank you for your love, support, and grace."

RB: "For those of you who take your coffee black, peppers hot, and unreferenced assurances with a grain of salt."

Table of Contents

About the Authors...xiii

Acknowledgments xv

Chapter 1: OI, AI, and Research or Why OI Is the GOAT and AI Is the BLOAT...1

The Mind of a Child…...2

Why Research Matters – And What the Heck Is Research Anyway?....................3

    Prove It 6

    Coffee Studies 10

Research in Design 12

Timing Is Everything 15

Recap 17

Before You Go…...18

Chapter 2: Neurocognitive Foundations for a Nonscientist (a.k.a. Brain Science, the Squishy Bits)......................................19

Engine of Thought – Parts Is Parts...19

Just a Jump to the Left…and a Step to the Right 23

Information for Humans 26

    It's All in Your Head...27

    Neuroplasticity 30

AI Has No Squishy Bits...31

    A Very Brief History Lesson 32

v

TABLE OF CONTENTS

What the Heck Is AI?..33

What AI Isn't 38

Recap 39

Before You Go…...40

Chapter 3: All the Feels 43

Why We Feel the Way We Feel 44

Emotions and Staying Alive...46

Evoking Emotion 48

Cognitive Biases...50

Attentional Bias 51

Endowment Effect 52

Halo Effect..52

Observer-Expectancy Effect 53

Primacy Effect 54

Recency Bias 55

The Dark Side: Patterns That Are Used to Manipulate 56

False Urgency...58

Basket Sneaking...60

Confirm shaming 60

Forced Action...61

Subscription Trap..62

Interface Interference...64

Bait and Switch 65

Drip Pricing...67

Disguised Advertisements 69

Nagging 70

Trick Questions 73

TABLE OF CONTENTS

SAAS Billing 74

Rogue Malware 74

The Dark Side and AI 75

Recap 76

Before You Go… 77

Chapter 4: Being Part of Something 79

What Does It Mean to Belong? Evolutionary Foundations of Human Groups 79

Mirror Neurons and the Art of Reflection 83

The Psychology of Belonging 84

Let's Play the Belonging Game 87

Power of Play 87

Gamify All the Things (or Not) 92

Designing for Trust 96

Nurturing Belonging Through Community 98

Cults vs. Communities: The Dark Side of Belonging 99

Recap 102

Before You Go… 103

Chapter 5: Defining the Box 105

Understanding the Limitations of Human Experience 105

Sensation + Perception = Experience 107

Sensation and a Preference for Detecting Change 110

Time Travel: Perception and Time 111

AI Making Things Easier…or Not? 116

Working In and Out of Limits 118

Forgetting Limits: Rage Clicking 119

Recap 123

Before You Go… 124

vii

TABLE OF CONTENTS

Chapter 6: Attention (or Lack Thereof) 125

The Bottleneck and the Pie 126

Distractions: What Were We Talking About? 131

How Much Is Too Much: Cognitive Limitations 131

I Hear What You're Seeing: Competing Inputs 132

Business As Usual: Habits and Routines 135

Designing for Routines 137

Designing for Habits 140

The Mythical Multitasking Beast...146

Old Problem; New Wrapper 146

Whither Hast Thou Gone the Yeti, Unicorn, and Multitasker? 149

Context Switching 152

Designing for Attention: Make Hammers Not Heroin 152

Recap 157

Before You Go... 158

Chapter 7: The Evolution and Revolution of People 161

Back in My Day: Age-Related Design Considerations 162

IYKYK: Expertise-Related Considerations..................................165

Cultural Considerations..169

The "We" in West Should Be Replaced with an "I"...................175

Uncertainty Avoidance and Its Impact on Evaluating New Products 178

The Yin and Yang of Purchasing Decisions 179

Importance of Personas (or Are They Personae?).........................180

Creating and Using Personas: The Right Way 181

Recap 185

Before You Go... 185

viii

TABLE OF CONTENTS

Chapter 8: Communication Is Hard (and We Suck at It) 187

Language: From Signals to Information 188

Production vs. Comprehension 192

Trigger Warning: Words As Emotional Cues 194

What You Heard Is Not What I Meant 197

Design Systems As Language 198

    Effective Use of Design Systems 199

Storytelling 202

    Improving Understanding 204

    Enhancing Engagement 205

    Improving Storage 205

Recap 206

Before You Go... 207

Chapter 9: I Remember When...or Do I? 209

Memories Come in Different Flavors... 210

...And Have Different Expiration Dates 216

Working Memory: Blood, Sweat, Tears, and...Neurotransmission 220

    Making Memorable Designs 224

Cognitive Cartography (and Other Alliterative Matters) 227

    Designing with Memory Maps in Mind 230

Remembering Some Cognitive Biases 233

    Recognition over Recall 236

Recap 238

Before You Go... 239

ix

TABLE OF CONTENTS

Chapter 10: Making Decisions: Why We Buy Lottery Tickets 243

How Numbers Stymie (Some of) Us 244

Data and Design: Don't Go with Your Gut...255

Is the Juice Worth the Squeeze?..258

Decisions Under Certainty vs. Uncertainty 261

Scarcity and Abundance: Impacts on Decision-Making 265

Foraging for Information...268

Influencing Perception and Persuasion...269

Foot in the Door 270

Door in the Face 271

Low Ball...272

Something for Nothing 274

Recap 275

Before You Go...276

Chapter 11: Learning (and Making Mistakes) 279

"You'll Learn to Love it": Why Understanding How Learning Works Is
Important for Design 280

Associative Learning...282

The Carrot Is Mightier Than the Stick...285

"Learning Scientists Swear by This One Trick..."..290

Why We Suck at Remembering Some Things (and What We Can
Do About It) 294

Failing Forward 297

Recap 299

Before You Go...299

x

TABLE OF CONTENTS

Chapter 12: Business, Research, and Design Relationships: It's Complicated 301

Design Thinking: No Pipe Cleaners Version 302

Empathize – with Data 303

Define: Problems Before Solutions 311

Ideate: Finally – Solutions! 320

Prototype: Try Before You Buy 322

Test: Measure Twice, Cut Once 322

Design Thinking: Are All the Sticky Notes Worth It? 326

Timing Isn't Everything – It's the Only Thing 326

What Should I Ask When? 327

Timely Research Is Timely 331

Recap 332

Before You Go… 333

Chapter 13: The AI Elephant in the Room 335

Different Timetables: Moore's law vs. Evolutionary Change 336

Self-Correcting…or Not 338

Bias In, Bias Out 339

Personification of All the Things 340

Soft Skills Are Hard 343

AI and (Not So) Soft Skills 346

Utensils Not Users 347

Recap 349

Before You Go… 350

Index 351

xi

About the Authors

Dr. Jerome L. Rekart has spent his career trying to convince anyone who will listen that by studying the brain and behavior we gain access to the user manual for what it means to be a human. He is a professor in the Department of Liberal Arts and Sciences at the Berklee College of Music and founder and principal of a research consultancy that helps established organizations and startups understand their users better. In previous iterations, he led teams of researchers who worked with corporate and nonprofit stakeholders to measure and characterize the impact and value of innovative offerings and before that was an Associate Professor of Education and Psychology. He is the author of the book *The Cognitive Classroom* as well as articles examining topics ranging from employer thoughts on AI in the workplace to the impact of skills-based training for retail workers to the molecular basis of learning and memory. He earned his B.S. in Biochemistry from Indiana University and his M.Sc. and Ph.D. in Psychology from Northwestern University and did his postdoctoral training at the McGovern Institute for Brain Research at the Massachusetts Institute of Technology.

ABOUT THE AUTHORS

Dr. Rebecca Baker started her career as a research assistant at the Space Vacuum Epitaxy Center at the University of Houston doing cool stuff with science and moved into software design when she discovered physics didn't have enough people-y bits. She has authored a book, *Agile UX Storytelling: A zombie software case study* (available on Amazon, it makes an excellent gift), holds a patent for information encapsulation, and has spoken at a ridiculous number of conferences. A UX Jedi, excellent listener, 30-year veteran of the software design space, and dungeon master, Dr. Baker's publications and talks span topics from technical writing to remote usability testing to agile UX processes and beyond. Her passion for research and helping people understand the "why" behind design combined with the recent developments in large language model (LLM)-based artificial intelligence led her to partner with the marvelous Dr. Rekart to write this book.

Acknowledgments

Although our names are the ones on the spine, this book, as with any endeavor of this scope, is only possible through the help and work of many others.

The illustrations throughout the text are the work of the fantastic Anastasia Soroka, who was able to take roughly articulated descriptions like "dapper hippopotamus" or a "ship laden with rubber ducks" and somehow produce drawings that were at least 90% of the way there on the first try. Truly a pleasure to work with and we highly recommend her.

To everyone at Apress, we think you showed great wisdom in greenlighting our book, and we appreciate the space and flexibility you provided for our voices to shine through. You just "got it" and this book wouldn't be possible if you hadn't.

And finally, a special thanks to our families. Writing a book often doesn't look like work, yet it is and requires hours of time that could be spent with all of you. Thank you for your patience, understanding, and encouragement.

CHAPTER 1

OI, AI, and Research or Why OI Is the GOAT and AI Is the BLOAT

RB: That's a lot of abbreviations and acronyms. Do you think folks will think we're Gen Z or something? Will they know what this book is about?

JLR: I think they'll understand that original intelligence (OI) still bests artificial (AI) and that the advantage that human designers will always have over artificial intelligence is pretty straightforward: their humanity.

RB: And understanding those thought processes that are uniquely human is how we can create better and more effective designs.

JLR: We aren't saying that AI is terrible *per se*. And we aren't worried that it is going to be the end of us. But we think it is important to identify the ways that our humanity is important because we can empathize with users, who are presumably also human, and can understand some of the reasons why they might like what they do. What we plan on doing with this book is provide a manual that outlines how the brain, cognition, and design go hand in hand (in hand). We're also going to back up everything that we relate with real-life examples and evidence from research studies that will reinforce that there are general principles that can be followed to make the best design decisions possible.

In essence, this is a user manual for your users' minds.

© Jerome L. Rekart and Rebecca Baker 2025
J. L. Rekart and R. Baker, *Designing for Human Intelligence in an Artificial Intelligence World*,
https://doi.org/10.1007/979-8-8688-1418-1_1

CHAPTER 1   OI, AI, AND RESEARCH OR WHY OI IS THE GOAT AND AI IS THE BLOAT

The Mind of a Child...

We'd like to start off by first showcasing how amazing the human mind is given that it can picture scenarios, objects, and settings that have never and will never exist. For example, we recently "asked" several artificial intelligence image generators to produce a simple image with this prompt:

"Draw a horse riding an astronaut."

The responses we received varied from "I'm not able to generate such an image" to pictures (each from a different model) that ranged from cinematic to somewhat cartoony to inexplicably patriotic; however, for months and months, not one of these actually fulfilled the prompt. The images produced all had an astronaut (with a helmet) riding a horse (without a helmet...poor thing).

Indeed, it wasn't until 11 months later that we were able to (finally, sort of) see an image that approximated a horse at least sort of behind an astronaut (still, alas, without a helmet).

On the other hand, the first (and only) five-year-old ("Jack") we asked to draw this image generated this:

CHAPTER 1   OI, AI, AND RESEARCH OR WHY OI IS THE GOAT AND AI IS THE BLOAT

What can be seen from this image is that the horse, with its four vertical legs, is clearly situated above the two-legged astronaut. The fact that Jack's image stylistically is quite primitive (at least compared with those generated by various AI models) is not important. What matters is that a child could imagine it without any training, redirects, or other interventions (which likely took place over the months and months of time during which all of the AI models were failing to produce what was asked for).

The reason "why" our minds – even at an incredibly young age – can perform tasks like real imagination as well as so many other wonderful things will be explored in this book, along with the "so what" as it relates to design.

We'll also talk about what design is – because it isn't just art. Steve Jobs famously said "Design isn't just what something looks like and feels like. Design is how it works." And that's true – your interactions with designed objects whether software like TikTok, physical items like doors, or your favorite comfy chair, and more are not just a result of color choices and aesthetics but the behavior and construction of the object itself. A door that doesn't open can be pretty – but it is a pretty bad design.

Why Research Matters – And What the Heck Is Research Anyway?

Truth can be a tricky thing. For example, in 2015, people were adamant that their experience of seeing a dress as either "gold and white" or "black and blue"[1] was the correct one (the dress was black and blue in reality). The ability to remove personal experience and subjectivity from truth is critical and is where research shines, particularly when it comes to identifying what users feel, want, and experience, as well as human behavior.

---

[1] https://slate.com/technology/2017/04/heres-why-people-saw-the-dress-differently.html

CHAPTER 1   OI, AI, AND RESEARCH OR WHY OI IS THE GOAT AND AI IS THE BLOAT

It's a common misconception that talking to someone who uses the product/process you're trying to improve is the same as doing research. For many designers, it can be a novel and empowering experience – they may have spent their entire careers up to that point designing interfaces for people they've never met. And certainly, it's important to get that contact – to reinforce that you are not your user and be able to internalize those differences. BUT that's not research. And while it's good to do, without coupling conversations with research, it runs the risk of doing more harm than good.

You end up assuming everyone is like the one (or three) user you met for 30 minutes. You also start assuming they told you everything they know and that they have laser-focused and helpful insights that, frankly, they usually don't. Most users aren't sure what their problems are let alone how to solve them. But that's why coupling product direction and design with good research is so important. It can give you a level of confidence in your direction that you can't get any other way. And don't get us started on the famous C-suite "gut"!

And it's not just "do research." It's knowing what kind of research you need to do and why. It's not just a box you can check blindly. It needs to be thoughtful and targeted – otherwise, it can be misleading and just a plain waste of money. And – there are so many "ands" – knowing when you can trust your results. That last piece is the key.

There are many reasons why you may need to question results. For example, you may not have spoken or tested with enough people. There are times when so-called "guerrilla" testing works out and yields fine results, but knowing when you can utilize this sampling approach is the part that can be tricky. So-called "convenience samples" or "guerilla testing" can help with things that, at a high level, could be recognized by anyone who is a typically developed human, like "is button 1 bigger than button 2?", but for anything more complicated, you're going to need more than three or five (sigh) people.

CHAPTER 1   OI, AI, AND RESEARCH OR WHY OI IS THE GOAT AND AI IS THE BLOAT

Knowing just how many people to ask can be challenging. And many user researchers operate as one-person shops or are designers who conduct research for validation, so they don't have the opportunity to obtain feedback on their methods or get different perspectives. For professional scientists – regardless if they are in industry or academia – the peer review process is one avenue via which trust is built. This means that it isn't enough that you think that "X" causes "Y"; the data that you've collected have to be compelling enough to convince other experts in your field (i.e., your peers). And the methods you selected need to be deemed rigorous enough to support your claims.

This process of peer review is a critical foundation of the scientific process. What many individuals don't realize is that academic journals don't pay the authors and authors don't pay the journals! For those wondering about the socialist aspects of this, just know that the business model is such that academic institutions, libraries, etc. are the consumers who fund the printing and pay the publishers. The economics of that aside, the reality that published research is not a commercial endeavor for researchers is of critical importance. Thus, because the studies that have been cited or will be discussed in this book went through peer review, you can rest easy that they can be trusted.

Furthermore, peers who review studies and make determinations about whether or not they should be published don't receive monetary compensation either (often, they will receive free access to the journal for a month or so as a token of appreciation).

Thus, there is

1. No direct financial incentive for researchers to publish their results.

2. No financial incentive for peers to accept the results or buy into the findings or interpretations of authors who submit to journals.

CHAPTER 1   OI, AI, AND RESEARCH OR WHY OI IS THE GOAT AND AI IS THE BLOAT

3. And because peer reviews are anonymous –
    meaning that the authors of the papers don't know
    which experts or peers in the field are reviewing
    their work – there isn't any social or other incentive
    for reviewers to accept findings that aren't
    methodologically rigorous or have findings that
    meet acceptable levels of validity.

That said, it's important to make sure the study you're referencing (and the journal for that matter) adheres to these guidelines. It is not uncommon these days to come across "journals" that are little more than a front for special-interest group publications that are not peer-reviewed. Which means you need a bit more understanding of research so you can determine how trustworthy those sources are. So, results in journals can usually be trusted, but you should check them ("trust but verify").

Prove It

Now, it can be tricky once you've found results to know if they are applicable to your particular instance or use case. This speaks to the "generalizability" of the findings. And this is something that can trip out a lot of folks, because they assume that because everyone is unique, that every aspect of them must also be unique. Because each person has their own set of experiences, biology, etc., many people wonder how results from the literature or some research study can apply to what they're trying to design.

The answer to this query becomes easier to understand if we first segue into a discussion of the two main branches of inquiry that researchers use. The first, which is primarily used in the social sciences, is inductive. With inductive approaches, like interviews and focus groups, the aim is to find commonalities and themes that help to explain or flesh out a theory.

CHAPTER 1   OI, AI, AND RESEARCH OR WHY OI IS THE GOAT AND AI IS THE BLOAT

The only way to fully know if the theory is correct, however, is through deductive methods, such as testing. The interplay between these two methods then yields progressively greater information and understanding about a phenomenon.

So, with the example above, we could have a new website that we think is light-years more intuitive than what we had. So, you've got this new site, so how will you determine if it is in fact more intuitive?

First, we need to do what is called "operationalize" what we mean by "intuitive." This really just means define the thing that you want to study. For this example, what you really mean is "easy to navigate," which is still a little vague but is now getting us closer to tangible metrics that we can use to test your assumption. Given that "ease" is something that you are striving for, we know that the more difficult something is to navigate, the longer it takes, so we can operationalize "intuitive" by evaluating the time to complete various tasks.

Now that we've done that, we move into the next phase of this deductive approach, which is the actual testing. To test out our assumption that the new site is more intuitive than the old, we will have to determine the parameters that we will examine. And this is where things get fun.

We are going to try our hardest to design the study so that we ask our users to perform the most difficult tasks and work their way through the most complicated – but realistic – scripts possible. Not the easiest or the one that we just "know" is right. But the one that will tell us whether they (the users) actually understand how to get around. This is a key premise of hypothesis testing that is very often missed by nonscientists. Studies are conducted to disprove theories, hypotheses, and hunches – not to prove them.

In fact, scientists don't actually look to prove anything.

That's right. One way of looking at it is that research doesn't tell you what's definitely, positively, 100% true – it tells you what *isn't* true. That's why we get so excited when the hypothesis is wrong. Weird, right?

7

CHAPTER 1   OI, AI, AND RESEARCH OR WHY OI IS THE GOAT AND AI IS THE BLOAT

As people, we don't want to be wrong, so if we design a study to prove that we're correct, we can stack the deck in our own favor, which means that we won't know if something is actually working or better or more delightful.

Here's a classic example. When one of our (Dr. Rekart's) daughters, S, who is now (muffles her actual age) years old was seven years old, she was playing with her baby sister, J (third of three), who was about one at the time. After a little while, S said to him:

"Dad, look at this. J knows her name."

Intrigued by where this was going to go, he said "OK, show me."

With a beaming smile, S looked over at her baby sister, who at the time was engrossed with some plastic blocks, and said: "J," and sure enough, little J looked over at her sister and smiled.

"See, Dad. I told you."

Now, if he weren't a scientist, this story would be a cute anecdote but not really have a point in the context of this book. But let's zoom out for a minute and consider that S had a hypothesis. That hypothesis was that her sister recognized and responded to her own name; in other words, J knew that she was "J." S then set up a test of this hypothesis by trying to prove that J would respond, which she did.

If he weren't a scientist (or honest), Dr. Rekart would tell you that the next part of this story was fabricated for illustrative purposes. But he IS a scientist and although that may not help him win any Father of the Year awards, it will perfectly demonstrate the point about setting up your test to try and disprove your theory.

"Hmmmm. OK. Let's try something else," he said to S.

I want you to call her again, but this time, use a different name. I want you to use the same voice, but this time say, "Tiberius." S gave him an odd look, but did as she was instructed.

S called out in J's direction, using the same singsongy voice that people seem to use with small humans, "Ti-ber-i-us."

CHAPTER 1   OI, AI, AND RESEARCH OR WHY OI IS THE GOAT AND AI IS THE BLOAT

And, just as S called her sister's name, J turned toward her and smiled in response to Captain Kirk's middle name. Thus, rather than associating the word "J" with herself, she was likely just responding to sounds that were directed at her. Or to her sister in general. Or to the singsongy intonation.

Or not. Here's where the testing piece can tell us whether something is disproven or not but usually cannot provide the details outside of what was tested. So, while we can tell you that J at that point in time did not look only when her name was uttered (she has subsequently caught on), we cannot tell you why she turned her head.

And, to the earlier point, we can't generalize the results. Do all children at this age respond to *Star Trek* references if phrased appropriately? We can't say. Even if we added in data from a thousand children, we wouldn't be able to say WHY the child responded. We could only eliminate reasons. This underlines the importance of combining quantitative and qualitative studies – "numbers" and "people" in other words. Big number studies can point us at the areas that we have confusion around. Then we can talk to people and drill into the "why" behind those results.

A good example of this is a piece of workload software Dr. Baker was working on. They had big numbers that indicated that people were consistently running extra reports at a point in the process where things were still in flight. In other words, the work wasn't done, so the report would only tell you what was in process, which didn't make sense. Why would people do that?

So, she went on site, asked questions, and watched the workload administrators do their jobs. It turned out the inputs required to make this process run were so convoluted and confusing, that they had created physical checklists to ensure they completed all the steps. They were running the reports to make sure all the processes were in progress because waiting until the end would've been costly. It seems obvious in hindsight, but we needed all the data – the initial strange report-running behavior that raised a flag followed by the onsite observations to uncover the why to be able to identify the problem that was occurring.

9

CHAPTER 1   OI, AI, AND RESEARCH OR WHY OI IS THE GOAT AND AI IS THE BLOAT

In both of these examples, you'll notice we talk about a particular population or role that the user is playing (child at a particular developmental level, workload administrator). We're not trying to generalize results to the entirety of humanity but rather to how humanity in its infinite variety fits into a particular role or context. Context is key in being able to make your findings more broadly applicable. When you are buying something, for example, you fall into a particular role based on your context (wanting to buy a dress) that makes you more similar to other people who have the same context. Likewise, if you are running a 5K, you will be more similar to other people running a 5K during your registration experience than, say, people who are browsing movies on Netflix. That's not to say that designing experience for both may have similarities (browsing for goods whether they are movies or races is remarkably similar), but my ability to generalize your needs for a complete experience (from browsing to watching vs. running) varies based on your context.[2]

Coffee Studies

Keep in mind that we're not saying that *nothing* can be generalized. Rather, you need to consider the results from any single study in the context of the individuals who were tested or observed. This is why replication is so important for science. When a study is run with different individuals but provides the same general patterns of results, it starts to become evident that the phenomenon in question – whatever it is – is perhaps not affected by what makes those folks different from one another and may be applicable to other groups.

---

[2] AI kind of sucks at this. The way AI is programmed, it "thinks" that all context is the same. So whether you're buying toilet paper or life-saving medical treatments, it will treat the experience identically. Context is the great differentiator that humans introduce into the equation – and understanding how the brain works lets us lean into that superpower and design amazing experiences.

CHAPTER 1   OI, AI, AND RESEARCH OR WHY OI IS THE GOAT AND AI IS THE BLOAT

Case in point, when Dr. Rekart was in graduate school working on his doctorate (and thus not yet "Dr."), he drank a lot of coffee. He knows people joke about how much coffee they drank, but he was hitting eight cups a day. While performing some literature reviews on neurodegenerative diseases, he came across a news report that referenced an article that initially provided him with some cheery news about his brewed caffeine habit.

The article was published in the prestigious *Journal of the American Medical Association* ("JAMA"), and in it, the authors found that individuals who drank coffee had a lower incidence of Parkinson's disease than non-coffee drinkers. His initial thoughts after reading the news article were of elation and some relief – because he felt like he had the evidence to justify his excessive coffee drinking to himself and others who were starting to wonder if being so jittery was actually a good thing.

But after that initial feeling of redemption subsided, he realized that he should probably dig into the source material a little deeper. Once he read the actual JAMA article, he realized that the study was observational and conducted entirely with individuals of Japanese-American descent (which he is not), who lived in Hawaii (he was in Illinois). Given these differences, he carefully recalibrated whether he should look at his excessive coffee drinking as healthy or not.

Now, the paper that he read (Ross et al.) was published in 2000. Since then, numerous studies have replicated the observations of those authors, with different people, living in different regions of the earth, and of different ages. Thus, with replication, we now know, some 20 years later, that moderate, daily coffee drinking (though not eight cups a day) may lower the risk of Parkinson's disease.

Research from multiple studies validated the generalization of that initial study. Generalization is the key.

Generalizable findings are what we will be referencing as best practices in this book. The evidence that we cite may, with time, have a different interpretation, but the utility of the findings for design is likely to persist

CHAPTER 1   OI, AI, AND RESEARCH OR WHY OI IS THE GOAT AND AI IS THE BLOAT

because the underlying research has been validated and corroborated multiple times and, importantly, by different laboratories working with different people.

Staying with the coffee theme, another study was done associating drinking your coffee black with having sociopathic tendencies.[3] And I can assure you that, although we both take our coffee black, neither of us could be classified as a sociopath.

Probably. Anyway, the generalizability of a single study's results are usually questionable at best. A single study can give us a direction to look in and explore rather than an ironclad conclusion we can use to make our decisions. Just like those guerilla, three to five user studies.

Research in Design

So now, you may be feeling better about the cream and sugar in your coffee. But you may be wondering "How does this relate to design?" or "How are existing research findings usually used or incorporated into design in practice?"

And the answer is, like a Facebook relationship status, it's complicated.

Unfortunately, a lot of research used by design is…questionable. Either the research itself is questionable (too small a sample, misused methodologies, biased collection instruments, etc.) or the way the research is used is questionable (reasoning from the parts to the whole, overgeneralizing, etc.). Here's a great example of questionable research. Dr. Baker was interviewing at a company years ago and asked them about what type of research they did. They proudly told her that they did A/B testing and she was impressed! A/B testing is a great way to get preference data on small design changes. You split your traffic coming through your

---

[3] Sagioglou, C., & Greitemeyer, T. (2016). Individual differences in bitter taste preferences are associated with antisocial personality traits." *Appetite* 96, pp. 299–308.

12

CHAPTER 1   OI, AI, AND RESEARCH OR WHY OI IS THE GOAT AND AI IS THE BLOAT

system into two streams – one experiences the current design, and the other experiences the change you want to test. Then you see which stream performs better. You have to keep the change small to make sure you know what you're testing (e.g., if you change both the location and the color of a button, which change is responsible for the improvement?), and you have to have access to large numbers of users to ensure the accuracy of the effect. She immediately followed up asking what software they used to do the split traffic. The interviewer looked very confused. So, she dug a little deeper.

"How are you doing your A/B tests?", she asked.

He explained that they invited six users on site and showed them one set of screens and then showed them a second set of screens and asked which one they liked better.

That's not research – you wouldn't be able to confidently draw ANY conclusions from that data.

Worse, this company THOUGHT they had good reason to follow the "results" of this experiment and that it would be as robustly valid as if they had done actual A/B testing. As a result, they chose an entire workflow that cost a lot of money to implement but didn't improve anything.

Another example is misusing good research. This happens everywhere and is often perpetuated in social media. When Dr. Baker was pregnant with her first child, she was part of an online community of moms who were all due in the same month. After they all gave birth, they stayed together, sharing stories and tips as they navigated motherhood.

At one point, there was a huge divide between some of the moms on the benefits of breastfeeding. It was very tense, and each side of the debate had a lot of feelings around it. She decided to take a moment and trace back the actual research that had been done on the subject. It turns out that most of the "antiformula" arguments referenced a single research

13

CHAPTER 1   OI, AI, AND RESEARCH OR WHY OI IS THE GOAT AND AI IS THE BLOAT

paper.[4] She took the time to read it and was surprised by what she found. The study looked at women in third world countries that did not have access to clean water. They discovered that children who were breastfed got sick far less than children who were formula fed – largely due to the fact that the formula was in powder form and had to be mixed with the unclean water. The authors concluded that for this population and this type of formula, breastfeeding had better outcomes. However, every journalist that quoted the research had said that formula (any kind) made babies (everywhere) sick. Similar misuse happens fairly commonly in design – color preferences, optimal number of clicks, and so on.

Now, that's not to say that all design research is a mess. There are many good foundational studies that inform the basic heuristics of design. Things like ensuring system status visibility, emphasizing recognition over recall – these are solid foundations from cognitive science that we leverage in best practices across the board. And targeted research for products is also useful. Doing a real A/B study can inform simple design changes that will significantly alter site traffic and conversions. Doing usability tests can reveal points of friction in complex workflows that might have gone unnoticed. Doing cardsorts can uncover navigational problems and help design make informed decisions about hierarchy.[5]

In short, design uses research to provide the building blocks for the work (foundational) and the unbiased, outside feedback that refines the decisions made during the process. We often tell our teams that they are uniquely disqualified from being a user. They may empathize (really sympathize) with them through observation, but they themselves are never going to be a good user substitute due to the fact that they are

---

[4] Huffman, S. L., & Combest, C. (September 1990). Role of breast-feeding in the prevention and treatment of diarrhoea. *Journal of Diarrhoeal Disease Research*, *8*(3): 68–81.

[5] This is another place that AI falls down. AI cannot make good decisions about the generalizability of a research study. That is, it can find information but it cannot evaluate it. This comes back again to context.

CHAPTER 1   OI, AI, AND RESEARCH OR WHY OI IS THE GOAT AND AI IS THE BLOAT

product managers/designers/developers. They have too much insider knowledge to look at things in the way someone who is NOT a product manager/designer/developer sees them. They've taken the red pill and there's no way back.

Timing Is Everything

So, your next question may be "How do I know when I need to do research or incorporate it? Do I just need it all the time?"

This is a great question! The thing to ask yourself is this: What am I going to do with the answer when I get it. Remember how we talked about disproving your hypothesis? Well, a lot of the time, that is exactly what happens. If you've planned for that, then great! But more often than not, when researchers present their findings and they are not what the requester expected (spoiler alert! cognitive biases are real and annoyingly persistent!), they end up in a situation where they know what they're making is wrong, but it's too late to do anything about it. Lots of research in the field fails because the designer or the product manager or the CEO asks the wrong question or asks it at the wrong time, and results end up being thrown out because there are no actionable results due to timing or misalignment.

You shouldn't use research to confirm your understanding – you need to expect that the results will upend at least part of what you're trying to do and have enough runway to make use of that data. Always ask yourself "What question am I trying to answer?" and "What am I going to do with that answer?" If you don't have a plan to handle an answer that isn't what you expected, don't bother with the research – it will just be a waste of time and money.

And for goodness' sake, don't just assume there was a problem with the population/data! While that's always a possibility (and questions are good!), it is unlikely that just because it didn't match your expectations, the

CHAPTER 1   OI, AI, AND RESEARCH OR WHY OI IS THE GOAT AND AI IS THE BLOAT

data is wrong. A key part of any study is being open to not being correct. In fact, one of our favorite stories about this doesn't just illustrate what happens when the data don't go your way but when you have to rethink the entire study.

When he was a young researcher who had just started his own laboratory, Dr. Rekart's graduate mentor and memory researcher, Dr. Aryeh Routtenberg, was trying to train lizards (don't ask – they were cheap to obtain) to learn a maze. Turns out that even after weeks and weeks of attempting to teach the lizards a very simple maze (shaped like a T), neither he nor his lab assistant could get the lizards to do what they wanted. However, because they were attentive to all of the factors of the study and didn't take anything for granted, they did notice something peculiar.

The peculiar thing that they noted had nothing to do with the lizards but rather with the beetles that they were using as part of the training routine.

Because the beetles were bought in bulk (like at an insect Costco or InvertebrateMart), the ones that were used over the next few days were stored at room temperature and those that wouldn't be used until the following week were stored in the refrigerator (this slows them down and preserves them alive...sort of like carbonite for creepy crawlies). What the assistant noticed was that the beetles who came from the cold seemed to remember elements of the maze that their room temperature counterparts did not. Though this was certainly not what either the lab assistant or the professor had in mind, Dr. Routtenberg told Alloway to "forget the lizards and start training the bait (i.e., the beetles)." Thus, they scrapped the initial study and started exploring this new idea that somehow temperature was facilitating the memories of the beetles.

The findings of that study ended up getting published in the journal *Science*[6] and helped launch Dr. Routtenberg's career (historical note: he

---

[6] https://www.science.org/doi/10.1126/science.158.3804.1066

16

did not continue to study insects – or reptiles – after that but moved on to mammals, which are much easier to train and don't cause people to go "ewwww" when you describe your work).

Thus, you have to be open to whatever the results of the study are, as well as any factors that could be at work or influencing them. Though it is readily apparent from this last example that it is hard for a person to know what a lizard wants, at least enough to train them, it can be just as difficult to know what our users want, desire, or are affected by. This is why replication is so critical and also why the various topics and studies that are covered in this volume were selected, because they have been replicated and validated as having some generalizability beyond just one group of users, respondents, or test participants.

Confusingly, this means both (1) that you should trust in the results that we will relay and feel confident in considering them for your own designs and (2) that if you find that they don't do exactly what you planned you should consider why that is and whether there is something about the individuals you are designing for or the ways that you are assessing the implementation of the findings that is affecting the results.

But then again, the mantra of most researchers is that any good set of findings yields more questions than answers!

Recap

- AI is not the same as human (or original) intelligence. It's limited by the inputs available and incapable of true creativity and originality.

- Research is a rigorous, well-defined practice that requires careful thought and consideration to implement. Anecdotal observations of a few people are not the same as research and should not be given the same weight when making decisions.

CHAPTER 1   OI, AI, AND RESEARCH OR WHY OI IS THE GOAT AND AI IS THE BLOAT

- Understanding what makes good research can help you evaluate whether to trust a given source.

- Understanding what questions to ask and when is key when using research in business.

- Be flexible! By having a solid understanding of a given subject area and how research can/should be conducted, you can respond to (and investigate) surprising results.

Before You Go...

You may have heard the idea of an "alpha" in wolf packs – a strong leader to whom all of the other wolves defer. This concept is often used to describe aggressive male leaders in business "He's an alpha male, that's why he got the promotion." "His alpha tendencies attract the ladies." and to justify a behavior pattern of asserting dominance. But...is there such a thing? Turns out the alpha wolf myth (and yes, surprise, it is a myth) is based on a single study done in 1947 called "Expression Studies on Wolves" that was based on an incomplete interpretation of the observed wolf behavior, largely due to a lack of available subjects (a.k.a. wolves). Since that study, numerous authors have disproved this "fight for dominance" structure showing that wolf packs are organized as family units, not as aggressive, hierarchical, patriarchal groups. The author of the original study recanted his findings as soon as more observations were available – that is, he publicly and loudly said he was wrong, and now we have more information. What he did wasn't bad research – it was as good as could be done with the limited data available. Good researchers will always continue to question their results so they can test and extend their understanding of what we know. So, remember, what we know is "true" today may not be "true" tomorrow as we learn more.

CHAPTER 2

Neurocognitive Foundations for a Nonscientist (a.k.a. Brain Science, the Squishy Bits)

RB: So, Dr. R., how much brain science do I need to know? Is there a Wikipedia/CliffNotes/movie version I can use?

JR: So, I know the whole book can't be about this (can it?), so I think we can keep it somewhat high level without losing any of the coolness.

RB: LOL – technically, the whole book IS about it.

JR: Yes, I guess it is, isn't it <smiles ear to ear>.

Engine of Thought – Parts Is Parts

In terms of how much brain knowledge is necessary for design, we will hopefully cover the basics sufficiently so that the rest of the book will be easily digestible and...user-friendly. In order to do so, we first need to

© Jerome L. Rekart and Rebecca Baker 2025
J. L. Rekart and R. Baker, *Designing for Human Intelligence in an Artificial Intelligence World*,
https://doi.org/10.1007/979-8-8688-1418-1_2

CHAPTER 2   NEUROCOGNITIVE FOUNDATIONS FOR A NONSCIENTIST
            (A.K.A. BRAIN SCIENCE, THE SQUISHY BITS)

establish a useful metaphor for thinking about the brain. For our purposes, we should think of the brain as the *engine* of thought. In addition to being a not-too-bad name for a progressive rock band (appearing tonight only: Engine of Thought), it will be useful for future discussions because like the engine of a car, you don't have to know the specifics of how it works to use it to get to where you're going.

Furthermore, each component of a car's engine – the timing belt, crankshaft, spark plugs, pistons, etc. – has a specific function that on its own won't make the car go. Similarly, the brain has specific components – the four lobes, hippocampus, cerebellum, anterior cingulate, etc. – that are responsible for different elements of thought, perception, etc., but none of them operate on their own or are useful in isolation.

CHAPTER 2   NEUROCOGNITIVE FOUNDATIONS FOR A NONSCIENTIST
(A.K.A. BRAIN SCIENCE, THE SQUISHY BITS)

This is why cognitive scientists and neuroscientists continue to be stymied by the persistence of what are known as "neuromyths," which are commonly held fallacies about the brain and the way it works. For example, the old adage about "humans only use 10% of their brain" is really, egregiously, completely, and utterly wrong (not sure if we can completely convey how wrong it is any more strongly).

Any human who is only using 10% of their brain is technically brain-dead or in an irreversible coma. While there are fluctuations in the degree to which different parts of the brain are involved with a task (solving a problem, remembering a school dance, reading a blog, etc.), the relative increases in activity of one region do not mean that 90% of the rest of the brain is sitting idle but rather that there is a preferential activation and usage of those areas. Thus, even though you may read that the amygdala is "involved in fear" (which it is), this doesn't mean that (1) it is solely responsible for fear (it isn't) and that (2) other parts of the brain aren't also responsible for everything that goes into being afraid (they are).

Indeed, even "inactive" portions of a healthy brain are still always active and engage in what neuroscientists call "spontaneous activity" (note: that it is referred to in this way because we are still determining the underlying role it plays in brain health).

But, again, just like each component of an engine has an assigned role, so do various parts of the brain.

Before briefly describing the areas that we'll discuss at a broad level, it is important to reiterate that none of these functions occur in isolation – but rather if they are identified as being "responsible for," then they are *necessary* for that particular brain function to occur, but they are not *sufficient* (again – nothing works in isolation).

If we broadly think about the outside of the brain (see figure), we can see that there are four main divisions.

A: This is the **frontal lobe**, which is responsible for higher-order thinking, such as problem-solving, critical reasoning, decision-making, speech production, some long-term memory storage, and what are

21

CHAPTER 2   NEUROCOGNITIVE FOUNDATIONS FOR A NONSCIENTIST
(A.K.A. BRAIN SCIENCE, THE SQUISHY BITS)

referred to as motor skills (the control center for our voluntary muscles). In addition, much of what we think of as our "personality" is due to the activity of this region.

B: This is the **parietal lobe**. The parietal lobe is implicated in our sense of touch throughout the body (this region is actually directly behind – or what anatomists call dorsal to – the motor region of the frontal lobe), our sensation of pain, and our ability to differentiate between hot and cold. In addition, this region is responsible for our ability to make determinations about relative distance, how objects in space compare to one another, etc.

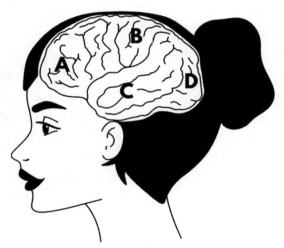

C: The **temporal lobe**, which is situated along the outer sides of the brain, is responsible for understanding language, the sense of hearing, processing sensory information, some basic emotional responses, and learning and short-term memory.

D: And, at the very back of the brain – at the farthest location possible from the eyes themselves – is the **occipital lobe**, which is responsible almost solely for our ability to see.

There are, of course, other regions of the brain (e.g., the cerebellum, which is located just beneath the occipital lobe), but their relationship to design is hard to identify, and it isn't really necessary to go into details here.

CHAPTER 2   NEUROCOGNITIVE FOUNDATIONS FOR A NONSCIENTIST
          (A.K.A. BRAIN SCIENCE, THE SQUISHY BITS)

In addition, it is important to remember that there are two of each cortex: one on the right side and the other on the left. So, in fact, you have two frontal cortices, two parietal, etc. We would refer to the two sides as the two hemispheres of the brain. And each hemisphere has each of the four lobes just described. And folks should know that "cortices" is the same thing as "lobes."[1]

Just a Jump to the Left…and a Step to the Right

Now, one thing to note about the two halves of the brain is that there is a fair amount of duplication of function. Thus, the right occipital lobe (or cortex) is responsible for the same visual functions as the left occipital lobe; the only difference between the two has to do with where the inputs come from – thus, for the left occipital lobe, sensation from the right eye (by way of the right thalamus) is processed there in the same way that sensations from the left eye are processed within the right occipital lobe. This crosswiring happens for a number of different parts of the brain, such as hearing and touch.

Before we move on, now is a great place to dispel another common "neuromyth"[2] that exists: that some people are "right-brained" and others are "left-brained."

The duplication of brain function just discussed is one reason – among others – why the idea that some individuals preferentially use only one side of their brain (e.g., "right-brained") is not actually based in fact. Indeed, a properly functioning brain requires a collaboration between

---

[1] Note that "lobe" is the same as "cortex." Neuroanatomists have a number of different, overlapping, and sometimes redundant ways of talking about the physical location of structures within the brain. Kind of like geographers and the earth. You have continents, but also hemispheres, etc.

[2] https://www.oecd.org/education/ceri/neuromyth6.htm

CHAPTER 2   NEUROCOGNITIVE FOUNDATIONS FOR A NONSCIENTIST
(A.K.A. BRAIN SCIENCE, THE SQUISHY BITS)

regions that are situated in both hemispheres. Furthermore, there isn't any evidence, using modern psychometric and imaging techniques from psychology and neuroscience, that supports the notion of "creativity" (though the operational definition is really important here) existing only one hemisphere, separate from mathematical or logical thinking.

Put differently, if you are a creative individual, then your brain is actually wired to perform problem-solving and use language – whether linguistic or mathematical – just fine. Indeed, correlational studies continue to show that highly intelligent individuals are also creative (and vice versa).[3]

And, though we may be belaboring this, the two hemispheres (right and left) are in constant communication with one another as a result of a massive number of connections that bridge the two, including the corpus callosum, the anterior commissure, and the hippocampal commissure.

In pioneering studies from the 1950s and 1960s,[4] Richard Sperry (who would later be awarded the Nobel Prize in Medicine for this work) and colleagues showed the impact of severing these connections in humans. These studies, which were carried out to provide relief to individuals suffering from severe forms of debilitating epilepsy, provided some of the first evidence that the sides of the brain *in some people* tend to specialize in specific types of tasks. These studies showed that tasks requiring language seemed to be solved best when information was presented to the left hemisphere and tasks requiring visual-spatial manipulations (think about mentally rotating a puzzle piece to see whether it will fit into a puzzle) seemed to be best solved by the right hemisphere.

---

[3] Silvia, P. J. (2015). Intelligence and creativity are pretty similar after all. *Educational psychology review, 27,* 599–606.

[4] Gazzaniga, M. S., Bogen, J. E., & Sperry, R. W. (1962). Some functional effects of sectioning the cerebral commissures in man. *Proceedings of the National Academy of Sciences, 48*(10), 1765–1769.

CHAPTER 2 NEUROCOGNITIVE FOUNDATIONS FOR A NONSCIENTIST (A.K.A. BRAIN SCIENCE, THE SQUISHY BITS)

Indeed, though Sperry and his colleagues (or others who replicated and continued such studies) never made assertions that either hemisphere was exclusively responsible for language or visual-spatial prowess, there were those in the public who took what were (and are) elegant findings and misinterpreted them through generalization, just like occurred with creativity and other functions of the brain.

For our purposes, what is most important is to know that the two sides are connected via a vast number of neural pathways, all of which are critical to normal functioning.

As we move within the brain, there are some structures (we refer to these as "within the cortex" or subcortical) that are also involved in some of what makes us human.

For example, the amygdala regulates a lot of what fear does to us, but contrary to common belief, it isn't responsible for our "feeling" of fear, which is due to a more distributed circuit throughout the brain. Suffice it to say that the amygdala does a fantastic job of modulating and influencing how emotion, like fear, impacts our memory.

It does this through interactions with the hippocampus, which is responsible for our ability to hold onto information and transform it into something that is portable and long-lasting: memory.

As anyone who has ever seen the movie *Memento* knows, when the hippocampus is damaged, people cannot form new lasting memories, yet they are able to maintain most of their old memories (before the damage). This form of anterograde amnesia occurs because the hippocampus isn't a "warehouse" *per se* of older information and memories but sorts and organizes information that is to be stored. When information is linked with emotions, like from the amygdala, this influences how the hippocampus regulates whether it will be stored, how long, and how vividly.

The hypothalamus is another important, subcortical structure within the brain that is involved in emotion. Neuroscientists will joke that depending on how crude you want to be, the hypothalamus is responsible for either the two Hs or the four Fs.

CHAPTER 2   NEUROCOGNITIVE FOUNDATIONS FOR A NONSCIENTIST
(A.K.A. BRAIN SCIENCE, THE SQUISHY BITS)

The two Hs in this regard are "homeostasis" and "hormones." The latter is fairly straightforward and has to do with signaling to various glands when to release different chemicals into the bloodstream that impact sexual responses, weight gain and loss, etc., and the former is about maintaining the status quo of what is required to keep a body functioning normally: like constant body temperature and the sleep–wake cycle.

For those more inclined to be a bit more raunchy, the four Fs (which is just a different way of remembering many of the same functions) encompass

- Feeding

- Fleeing

- Fighting

- Sex

Anyway, to round out the subcortical structures of note, we should mention that the thalamus (which is atop the hypothalamus) is a bit of a way station for sensory information. It receives information from our sensory organs (e.g., eyes, ears, etc.); does some initial processing, sorting, etc., of that input; and then makes connections with various cortical areas, like the occipital lobe, the auditory cortex (located in the temporal lobe), etc.

Thus, you can think of this as a series of different circuits, each of which organizes, adds, combines, etc. the information we receive from our senses and impacts the ways in which we perceive the world.

Information for Humans

So...what do we mean by information? To understand a bit more, let's consider Information Theory.

CHAPTER 2   NEUROCOGNITIVE FOUNDATIONS FOR A NONSCIENTIST
(A.K.A. BRAIN SCIENCE, THE SQUISHY BITS)

Information Theory, developed by Claude Shannon in 1948, is a mathematical theory that deals with defining and quantifying information as well as how it is transmitted and processed. It provides a framework for measuring the amount of information in a message, typically in the context of communication systems. Information, in this theory, refers to the reduction of uncertainty, and it's useful for us because it gives us a solid definition of what we mean by "information" and a lot of the words we use around information. Specifically, information is a measure of the reduction in uncertainty when a message is received. That is, how much this piece of data will make the overall meaning of something make more sense. It is quantified in terms of bits, where one bit represents the amount of information needed to distinguish between two equally likely outcomes – how many words will it take to help you understand what I meant. Overall, Shannon's theory provides a rigorous mathematical framework for understanding and optimizing communication systems and is used in many fields, including telecommunications, data compression, cryptography, and machine learning (spoiler alert – that's AI!).

How does that relate to information in the brain? Well, at its root, it's how we think about the signals the brain produces. There's a lot going on in there, and it takes a certain number of transmissions before what has been transmitted goes from "noise" (or as Shannon put it, entropy) to "information."

It's All in Your Head

Many people now realize, either from their own experiences or from television, that neurotransmitters are the chemicals found within the brain that are responsible for the conveyance of information. For example, serotonin, which is one of the neurotransmitters responsible for depression, can be increased and have substantial effects on whether someone is depressed or not.

27

CHAPTER 2   NEUROCOGNITIVE FOUNDATIONS FOR A NONSCIENTIST
(A.K.A. BRAIN SCIENCE, THE SQUISHY BITS)

Thus, chemical signaling, which takes place *between* brain cells, is mediated entirely by neurotransmitters. Serotonin, the levels of which are increased between brain cells by drugs like Prozac and Zoloft, is found at specific types of brain cells – but not all of them. This is because there are many different types of cells within the brain, many of which respond to some neurotransmitters but not all.

Parkinson's disease is an example of what happens when a particular set of cells that produce dopamine (another neurotransmitter chemical) deteriorate in the brain. Indeed, there are dozens of different types of neurotransmitters within the brain, and the impact of various drugs can preferentially target some parts of the brain differently than others.

This is why drugs like LSD produce particular types of hallucinations (due to interactions with serotonin) but cocaine (which impacts dopamine) largely does not. Similarly, some psychoactive substances, like alcohol, can impact many different neurotransmitter types, which can explain some of the different ways that low levels of inebriation affect people differently than being drunk and why some individuals get happy before becoming sad or angry.

So, you may be asking yourself, "But what causes a cell to release a neurotransmitter in order to communicate with other cells?" This is where the electrical nature of brain cells, which folks may also be familiar with, comes into play. When we encounter something in the outside world, like an image or a sound, the action of that external stimulus (i.e., the information) on our eyes or ears triggers an electrical response within those cells. Thus, when light shines on our retina, the rods and cones respond to it through an electrical change that releases neurotransmitters on the next cells in line (in this case "bipolar cells").

The neurotransmitters are released by the electrically active neuron into the gap between the cells – the synapse. The neurotransmitters that are released by axons then bind to the dendrites on the other side of the synapse of the receiving cells (axons release and dendrites receive). The binding of the neurotransmitter to the dendrite then creates an electrical

CHAPTER 2 NEUROCOGNITIVE FOUNDATIONS FOR A NONSCIENTIST
(A.K.A. BRAIN SCIENCE, THE SQUISHY BITS)

charge in the receiving cell, which will then cause it to either release its own neurotransmitters on the next cells in the circuit or will cause it to stop releasing neurotransmitters.

Thus, the information from the outside world is transformed (technically, it is trans*duced*, because it changes from one form of energy to another) from whatever form it is to an electrical charge, then a chemical messenger, then an electrical charge, then a chemical messenger, and so on, and so on.

Which sounds like a lot of processing. And just like making a copy of a copy, there is some loss of what the original message was. We need to remember that everything we think we are seeing, hearing, feeling, smelling, and tasting is really just the byproduct of something that our senses broke down into very basic signals and then our brain attempted to reproduce by putting everything back together as best as it could.

You can think of it in the following way.

Let's say you are given a fully constructed castle made out of LEGO (or, if you're really lucky, the Millennium Falcon). You are given the chance to examine it, and then it is taken away and disassembled and you are given the task of rebuilding it.

However, before you're able to reconstruct the original structure, some of those blocks are subjected to conditions that result in their distortion, like extreme heat, and some blocks are lost in the process (probably embedded in the underside of someone's bare foot).

Thus, though you'll be able to reconstruct – possibly – what you originally saw, it can never be a perfect recreation of the original. It just isn't possible. What's more, your interpretation of how to rebuild it is going to influence some of the three-dimensional elements that you couldn't see, couldn't remember, or just didn't care enough about to encode in the first place.

CHAPTER 2   NEUROCOGNITIVE FOUNDATIONS FOR A NONSCIENTIST
(A.K.A. BRAIN SCIENCE, THE SQUISHY BITS)

Neuroplasticity

Pretending for a moment that the average reader doesn't know what neuroplasticity is, we can think of neuroplasticity as the blanket term – and admittedly not a great one – to describe how our brain adapts and changes in response to external stimuli. Though the term "plasticity" is used here, one would be equally well served by thinking of it as flexibility or elasticity.

There are parts of our brains, like those involved in very basic functioning, that don't change over the course of our life. For example, cells involved in signaling messages from our brain (in this case, the medulla) to our heart aren't considered to be very "plastic," meaning that they don't change much based on the demands of external stimuli.

On the other hand, parts of our brain responsible for learning and memory are incredibly "plastic." This characteristic – the plasticity of cells in many of our lobes and subcortical regions – can take many forms and is an adaptive process that makes it so that we not only continue to learn new skills over the course of our life but also ensure that once we've mastered the basics of something, it will take less and less effort to expand upon that skill.

As anyone who has ever learned an instrument, like guitar, knows, the first stages that involve learning where to place your fingers on the strings, how hard to press, etc. all take what seems like an incredibly long period of time.

So. Very. Long.

But, with continued practice and diligence, the skill becomes more and more automatic. Learning your seventh chord takes much less time than learning your first (which was probably a G-major).

This is all due to neural plasticity in the motor cortex, neostriatum, hippocampus, and other regions across the brain.

Similarly, knowledge that is accumulated is actually stored through changes that take place within brain cells. This is why studying – which is really just the act of practicing with information – is critical to memory. It is through the act of repeatedly using brain circuits that you reinforce the importance of those signals.

CHAPTER 2   NEUROCOGNITIVE FOUNDATIONS FOR A NONSCIENTIST
(A.K.A. BRAIN SCIENCE, THE SQUISHY BITS)

And plasticity can take any number of different forms in brain cells. It can result in

- Restructuring of existing axons and/or dendrites

- Growth of new axons or dendrites

- Increases or decreases in the number of neurotransmitter molecules released from axons

- Increases or decreases in the amount of electrical charge that results when neurotransmitters bind to dendrites

- Other metabolic changes that impact the efficiency of brain cells

Suffice it to say that THERE is a lot going on in your brain at any given time.

And across all of it – not just 10%.

AI Has No Squishy Bits

"Artificial intelligence will save us all – no more busy work!"

"Artificial intelligence will bring the end of the world!"

"Learn how to use artificial intelligence in your business!"

The headlines in 2023 and 2024 are filled with praise for and imprecations against AI. But...what is it? Ask most people outside of the data science profession, and they will give you a vague, hand-wavy descriptive mix of facts, *Star Trek*, and anthropomorphism. So, let's take a minute to really understand AI.

CHAPTER 2   NEUROCOGNITIVE FOUNDATIONS FOR A NONSCIENTIST
(A.K.A. BRAIN SCIENCE, THE SQUISHY BITS)

A Very Brief History Lesson

Long before people started using ChatGPT to write performance reviews and letters to mom, artificial intelligence had a place in our hearts and our fables. Many mythologies include stories of intelligent automata, such as Talos, a giant bronze automaton created to protect Europa from pirates (and *Jason and the Argonauts*) by hurling giant boulders at incoming ships. Sort of the original security bot – a theme found in many science fiction works from *Star Trek*'s Data to the Murderbot in *The Murderbot Diaries*. Our work to create automated intelligence did not end with stories and shows. From the Mechanical Turk to Deep Blue, we have attempted to create machines that would match our intelligence – interestingly most often to play games.

The term "artificial intelligence" was first coined by John McCarthy in the mid-1950s. However, AI as we know it has its roots as early as the 1940s in work done on artificial neurons by Warren McCulloch and Walter Pitts and later by Alan Turing, as he and John von Neumann moved computing to binary logic, paving the way for much more powerful computing. In Turing's article "Computing Machinery and Intelligence," he begins to call into question the difference between human intelligence (or as we like to call it original intelligence) and computer or machine intelligence. Work and interest in artificial intelligence has continued through the years, accelerating with the advent of microprocessors in the late 1970s and more recently with the access to large data sets and powerful processors around 2010. From there, the development of large language models (LLMs) like Alibaba and BERT in 2017 paved the way for the AI we know and love (or hate) today.

But whether you believe that AI was the Engine predicted in *Gulliver's Travels* as a way in which "the most ignorant Person at a reasonable Charge, and with a little bodily Labour, may write Books in Philosophy, Poetry, Politicks, Law, Mathematicks, and Theology, with the least

CHAPTER 2   NEUROCOGNITIVE FOUNDATIONS FOR A NONSCIENTIST
         (A.K.A. BRAIN SCIENCE, THE SQUISHY BITS)

Assistance from Genius or study" or if you see it as the helpful and humorous Dr. Theopolis who assisted Buck Rogers in *Buck Rogers in the 25th Century*, AI has and will continue to permeate our everyday lives, making it critical to understand its strengths and weaknesses – as well as our own.

What the Heck Is AI?

Our current (2025) definition of AI usually refers to a type of AI known as large language models (LLMs). But there are many types and forms of AI that have been developed over the years.

Machine Learning

Machine learning is a set of algorithms that can learn new patterns and adjust through repeated use. There are algorithms for classification, regression, clustering, and more – each with its own strengths and weaknesses. You feed these algorithms data to complete a given task, and the algorithms look for patterns and relationships in the data that will help them make predictions or decisions. Through a process of trial and error, they adjust parameters until they find the best solution. For example, you might write a set of algorithms to identify dogs. You test them by showing the algorithms pictures of dogs, pigs, loaves of bread, and other things, identifying each one as a dog or not a dog. The algorithms identify patterns of dog vs. not dog, until they are able to reliably identify dogs. After that, they are able to generalize to dogs they have not seen before, in different settings or poses.[5]

---

[5] The Mitchell's vs. The Machines, Netflix Original Film, 2021.

CHAPTER 2   NEUROCOGNITIVE FOUNDATIONS FOR A NONSCIENTIST
(A.K.A. BRAIN SCIENCE, THE SQUISHY BITS)

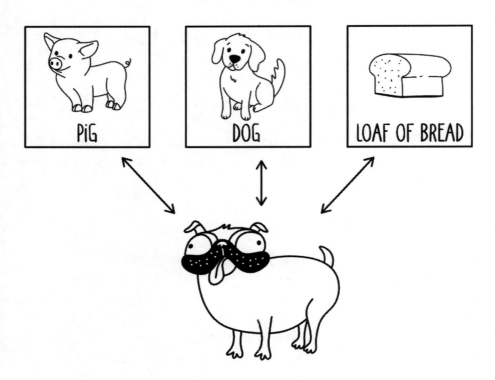

Machine learning algorithms can handle huge volumes of data very quickly, letting them tackle tasks that would be impossible or impractical for humans to do manually. And the more data you provide, the better the algorithms get. They are powerful but ultimately limited by both the data that they have been provided and the original directive.

Neural Networks

Neural networks are a complex web of interconnected nodes inspired by the human brain, used to process information in a nonlinear fashion. These nodes process information, each refining the network's understanding of the subject matter it has been tasked with – mathematics,

CHAPTER 2 NEUROCOGNITIVE FOUNDATIONS FOR A NONSCIENTIST
(A.K.A. BRAIN SCIENCE, THE SQUISHY BITS)

language, stock predictions, etc. Think of the network as a series of layers, each comprising virtual neurons. These neurons are connected to neurons in adjacent layers, forming a network of connections. When you feed data into the network, it undergoes a journey through these layers, with each neuron performing a calculation based on its input and producing an output.

Through a process called training, the network adjusts the strength of connections between nodes, gradually improving its ability to predict and generate results. By comparing its outputs to the desired outcomes and tweaking its parameters accordingly, the network learns to make better predictions and decisions over time.

Neural networks come in many shapes and sizes, each suited to different types of tasks, from recognizing faces in photos to translating languages or playing chess. They power self-driving cars and provide personalized recommendations on Netflix. However, they have their challenges – they require vast amounts of data to train effectively and are not foolproof. Designing and fine-tuning these networks can be as much art as science, requiring creativity and intuition to achieve optimal results.

Large Language Models (LLM)

Large language models (LLMs) like ChatGPT are the latest in a long line of AI developments. LLMs use huge systems of neural networks to understand and generate humanlike text based on large pools of data (sometimes called data lakes). At their core, LLMs are built upon a foundation of data and algorithms. Using books, articles, websites, and more, they identify the patterns and structures that make up human communication and then mimic them. When prompted with a question or topic, the model activates relevant nodes in its network, setting off a cascade of computations. These computations weigh the probabilities of different words and phrases, crafting a response.

CHAPTER 2   NEUROCOGNITIVE FOUNDATIONS FOR A NONSCIENTIST
(A.K.A. BRAIN SCIENCE, THE SQUISHY BITS)

What makes LLMs striking (and a little intimidating) is their adaptability. These models can pivot between genres, languages, and writing styles easily, assuming that those patterns exist within its training set. So, whether you're asking for a technical explanation, a whimsical tale, or an essay, the LLM can tailor its output to suit your query. However, while they excel at mimicking human language, they lack true understanding of consciousness. They also can come in "supervised" and "unsupervised" forms, which at their most basic level have to do with how much say the humans have in how the training goes (which feels very much like a discussion one of us had with his younger students about "kids today" and playing outside vs. when he was a kid).

Given the lack of consciousness (and, in some cases, supervision), they may stumble over ambiguous phrases or produce nonsensical responses when pushed beyond their training data. An example of this is in academic writing – if asked to create an essay with references, LLMs will produce an essay with references – except those references will be made up. The LLM understands what a "reference" looks like, but it lacks the ability to comb the literature, identify relevant publications, and then cite that as part of a synthesis. It lacks the ability to access a context outside its training data set.

In addition, because they rely on the availability of information (whether good or bad) as a marker of whether something is "known," they are unable to understand whether the information is accurate or not. For example, learning scientists, cognitive psychologists, and many educators have been trying (unsuccessfully) to stop the persistence of the idea of "learning styles" (like right brain/left brain) for decades (see Rekart & Heyman (2024) for review[6]); these ideas continue to be taught

---

[6] Rekart, J.L. & Heyman, J. (2024). Leveraging the Economics of Belief to Provide Pre-Service Teachers with Evidence-Based Alternatives to Learning Styles. *Transforming Teacher Preparation Through Identity, Development, and Effective Technologies.* Hershey, PA: IGI Global.

CHAPTER 2   NEUROCOGNITIVE FOUNDATIONS FOR A NONSCIENTIST
(A.K.A. BRAIN SCIENCE, THE SQUISHY BITS)

at some colleges, and thousands of schoolchildren are subjected to so-called "learning styles tests" that have as much validity as Buzzfeed quizzes related to "Which Harry Potter Character Are You?" or "Which Journey Song Describes Your Life?" Despite articles in peer-reviewed journals showing that these tests and their underlying theory are not only false, they may be harmful, the majority of links or hits on the main pages of search engines (which would be combed through by AI) will show the opposite. Thus, when queried "provide a summary of how I could find out my learning style," ChatGPT provided the following response to Dr. Rekart:

> "Certainly! Discovering your learning style can help you tailor your study habits and improve your learning experience. Here are some ways to find out your learning style:
>
> 1. Self-Assessment Quizzes:
>
>   i. Take online quizzes that assess your learning preferences. These quizzes often categorize learning styles into visual, auditory, kinesthetic, logical, social, solo, or natural. Some popular quizzes include..."

You can see that the system happily...and erroneously...provided him with what he asked for without any caveats or indication that there are issues with this. Indeed, there was even a helpful reminder at the end:

> "Remember that everyone has a unique blend of learning styles, and it's essential to adapt your approach based on what works best for you!"

Wow. So. Very. Helpful. (sigh)

CHAPTER 2   NEUROCOGNITIVE FOUNDATIONS FOR A NONSCIENTIST
            (A.K.A. BRAIN SCIENCE, THE SQUISHY BITS)

What AI Isn't

Despite how amazing and clever AI may appear, AI is not human. It lacks the motivation, emotion, awareness, and most importantly context that a person (even a newly minted person such as our five-year-old artist, Jack, from the previous chapter) has. AI may appear to be clever, creative, and even to show feelings, but it is none of these things. It is, at its heart, a calculator – a very powerful and useful calculator, but a calculator nonetheless. And as such, it is important not to personify AI, because that way lies misuse and ultimately disappointment. It is unlikely a malicious AI will take over the world vis-à-vis SkyNet (however, if your name is John Connor, maybe brush up on your survival skills just in case). It is far more likely that we will automate something that requires a nuanced, context-driven judgment, in the mistaken impression that AI has those capabilities, and suffer the consequences when it makes a decision that needs context, like the predictive policing seen in *Minority Report*.

CHAPTER 2   NEUROCOGNITIVE FOUNDATIONS FOR A NONSCIENTIST
(A.K.A. BRAIN SCIENCE, THE SQUISHY BITS)

For example, imagine an AI that has been put in charge of dispensing medical resources, such as medications, life-supporting machines, and bed space. Now imagine someone who has a terminal condition entering a hospital using that AI. Will the AI be able to recognize that it's important to help the person stay comfortable and help them evaluate their quality of life? Or will the AI evaluate whether the person should consume the resources available based on their life expectancy? AI lacks the ability to evaluate ethical and psychological contexts and complexities. It excels at efficiency and complex calculations – and there are many situations in which taking the most efficient route would end in tragedy.

Of course, this is also a route that humans can take as well. If we don't take the time to understand how our brains work and how we interact with one another, we risk making the same mistakes that AI does. The difference is that with OI, we have a choice.

Recap

- The brain is a heterogenous "engine" that has various parts that specialize in some tasks, but no one region is sufficient to carry out just that function.

- The two halves of the brain (right and left) work together seamlessly thanks to billions of connections between the two.

- Throughout our lives, our ability to learn and adapt is mediated by changes in the strength, number, and distribution of the connections within and across the brain.

- Artificial intelligence (AI) as a concept has been around for a long time.

CHAPTER 2   NEUROCOGNITIVE FOUNDATIONS FOR A NONSCIENTIST
             (A.K.A. BRAIN SCIENCE, THE SQUISHY BITS)

- AI is a category of different types of algorithmic processing such as machine learning, neural networks, and large language models.

- AI has a lot of potential, but it is neither going to save us or bury us on its own – it's just a model (like Camelot in Monty Python's Holy Grail).

- Humans need to make the effort to understand context to avoid making the same mistakes as AI.

Before You Go…

While the Mechanical Turk and Deep Blue may sound like punk bands, they were actually attempts at creating machines that could play chess – an ability we often associate with intelligence. The Mechanical Turk is one of the first documented cases of a chess machine. Wolfgang von Kempelen created the Turk in 1770 to get the attention of the Empress of Austria with whom he was smitten. A large wooden cabinet containing an intricate array of gears and topped with the torso and turbaned head of a mustached male, the Turk would be wheeled into court, and the cabinet would be opened to display the gears for the inspection of the onlookers. Once satisfied, the cabinet would be closed, an opponent chosen, and the game would begin. The Turk would more often than not win the game, smoke curling occasionally from its turban as it considered each move. The machine made the rounds for almost 84 years, defeating the likes of Benjamin Franklin and Napoleon Bonaparte, but eventually was revealed to be fraud – run by a short man who occasionally needed to light a candle to see (explaining the smoke).

40

CHAPTER 2   NEUROCOGNITIVE FOUNDATIONS FOR A NONSCIENTIST
             (A.K.A. BRAIN SCIENCE, THE SQUISHY BITS)

A complete fraud – so not an AI – but this illustrates our willingness to take seemingly miraculous mechanical feats at face value. It wasn't until Deep Blue, a true AI system, won against Garry Kasparov (the world chess champion at that time) in 1997 we started to see real strides in creating systems that could handle the complexities of games. Deep Blue was created by IBM (originally called Deep Thought – obviously the IBM programmers were fans of Douglas Adams' *Hitchhiker's Guide to the Galaxy*, as they should be) and is considered to be a major milestone in AI development.

CHAPTER 3

All the Feels

RB: So, our readers may be thinking. "But design is about creativity! It's about feeling! That's not something you can really quantify, right? Right?"

JLR: Well, as we both agree, actually you can quantify creativity...and most other human behaviors, thoughts, and feelings. We know, after all, when we are angrier, happier, sadder, etc. The "-ers" in each of those words indicates that there is more of something, and for there to be "more," it means that somehow we knew what there was and that, now, in our "X-er" state, we have more of "X."

RB: And design can be used to make us feel more or less of these feelings, which in turn can lead us to spend more time on a site or make purchasing decisions...even buying things we may not otherwise want or need.

JLR: This is not a new phenomenon but actually goes back over a century to the early intersection of psychology and advertising.[1]

RB: Yes, as we'll discuss, there are any number of ways that our emotions and motivations are manipulated through the psychological biases, heuristics, and phenomena that we'll discuss. So stay tuned...

JLR: Oh, I see what you did there. Nice.

---

[1] Kuna D. P. (1976). The concept of suggestion in the early history of advertising psychology. *Journal of the history of the behavioral sciences, 12*(4), 347–353.

© Jerome L. Rekart and Rebecca Baker 2025
J. L. Rekart and R. Baker, *Designing for Human Intelligence in an Artificial Intelligence World*,
https://doi.org/10.1007/979-8-8688-1418-1_3

CHAPTER 3   ALL THE FEELS

Why We Feel the Way We Feel

All of our emotions are the result of neurotransmitters being released at particular synapses in particular parts of our brain. For example, we know that the anticipation of something that is rewarding, like the bite of a cookie, releases dopamine all along the pleasure pathways that extend throughout our frontal cortex. Now, if you are a well-adjusted human, then you would likely have more dopamine release when anticipating the first bite of a chocolate chip cookie than what would be released when anticipating biting into an oatmeal raisin cookie (this is because oatmeal raisin cookies are the liars of dessertdom – they string you along with the thought that there is chocolate, but then you find a dried grape instead…seriously, who thinks that is good?). It is this differential release of neurotransmitters (in this case, dopamine) that causes the differential level of joy. Thus, feelings *can* be measured.

And this also illustrates exactly how feelings are evoked. Now, something like seeing (or smelling) a cookie is a fairly straightforward stimulus–response pairing for joy. However, other emotions/feelings that we experience are actually the result of a complicated interplay of the stimulus itself (e.g., the cookie), the external context in which we're experiencing the stimulus (e.g., Did we just have a big meal? Are we in a place that we normally wouldn't consider eating?), our internal context (e.g., Are we in the middle of a spat with someone? Are we stressed about our weight or sugar levels? Are we on a diet and this will cause a joy–shame stress spiral?), and the stored histories associated with all three (i.e., the stimulus and external and internal contexts[2]).

So, given all of that, we can see why feelings are, well, complicated. And yet so incredibly powerful that, despite it all, there are still some stimuli that are so universal that they almost never fail. Think about television ads that make everyone go "awwwwwww."

---

[2] Quintana, D. S., & Guastella, A. J. (2020). An allostatic theory of oxytocin. *Trends in cognitive sciences*, *24*(7), 515–528.

CHAPTER 3   ALL THE FEELS

Things like

- Acts of kindness
- Baby humans
- Baby animals: puppies and kittens are usually winners, although for our money we'd go with a baby hedgehog any day (which you could call a "hoglet" or, if you're feeling Dickensian, an "urchin")

Indeed, these types of images evoke what are called "prosocial" behaviors, which are responses that work to the betterment of the group and its members. Put differently, rare is the occasion when someone sees a puppy playfully licking the face of its owner and promptly thinks to himself, "I feel like getting into a fight." And we're willing to bet that nobody ever got to brawlin' after seeing a hoglet.

CHAPTER 3   ALL THE FEELS

Emotions and Staying Alive

The world is dangerous, and humans actually have a large number of disadvantages in the wild. To wit, *Homo sapiens*

- Is bipedal (i.e., stands upright on two feet), which is awkward and relatively easy to put off balance

- Lacks strong and noteworthy claws/talons, horns, or spikes

- Lacks poison, venom, or other chemical defenses

- Lacks pronounced teeth that could be used to tear apart a living prey

- Has a relatively long gestation period and low number of offspring per birth

- Is entirely dependent on its mother for the first nine to ten months of life

- Is largely dependent on a caregiver for the first one to ten years of life

- Is largely devoid of hair and has no fur or adult brown fat, meaning that survival in cold climes cannot occur naturally

Given all of those factors – and wow, we're not feeling so good about being people right now – if humans didn't evolve other means of surviving, they wouldn't have persisted as a species. So, how come we're all still around? Well, what we do have is

CHAPTER 3   ALL THE FEELS

- Well-developed binocular and three-dimensional vision

- A prehensile thumb, which operates differently than the other four digits of the hand, facilitating grasping and fine motor operations

- Social structures that facilitate teamwork

Now, you may be asking yourself why we didn't list something like "intelligence" or the "ability to write a book and use words like 'prehensile,'" and that is because the first three are likely responsible for the latter. That is to say that there were selective pressures (i.e., external factors that made some adaptations more useful) that facilitated the coevolution of high acuity vision, grasping dexterity, and the need to live in social enclaves. Taken together, these three likely drove what we now think of as intelligence, as those individuals who were able to do the first three likely had well-developed cortical regions, which facilitated their survival (and the longer one could survive, the more opportunities to pass on the genes responsible for those adaptations).

Emotions are the byproduct of that last part: the social structures. Relationships are complicated, so there has to be processes in place that allow individuals, who have their own wants and needs, to get along, to be able to accommodate and appreciate one another, and what we think of as emotions are the likely byproduct of mutations that occurred. Individuals with those mutations would have been better situated to live in groups and thus would have been more likely to survive and more likely to mate (i.e., more attractive), thus passing on their genes to create the next generation. And so on. And so on.

And that is likely the reason why we have feelings.

CHAPTER 3   ALL THE FEELS

Evoking Emotion

So you're probably wondering "What do I do with this emotion stuff?"

What can't you do with it? We can use design to lead people to a particular emotion – which then can reinforce what they want to do in both good and bad ways. In design we talk about "dark patterns" – these are design elements that are used to manipulate the user into doing something they might not actually want to do. We've all seen them. For example, you might be on a shopping site, and there's a little indicator that says "Only five left!" You're going to feel anxious, because that item might be gone if you wait too long, and that might lead you to buy something without fully considering whether you need or want it. This is a very simple example of using a design element to manipulate your emotions. Pictures of small fuzzy animals[3] evoke a feeling of happiness and can make you more open to the messaging that follows. Or badges that appear on the icons on your phone that make you feel as if there is something that needs to be addressed or done, so you click on it, which in turn raises your engagement with the application.

---

[3] Like baby hedgehogs/hoglets/urchins...

48

CHAPTER 3   ALL THE FEELS

Using design to influence your behavior can be good. Providing information about what is going on in the background can help you be more patient and less frustrated with processing times, lowering your stress. Providing celebrations when tasks are completed can instill a sense of joy and make you more likely to finish more tasks in a learning or work environment and facilitate your enjoyment while doing it.

And, this will allow us to get our desired result and build long-lasting relationships with our users. Manipulating people in a bad way always backfires – users will think twice about using your site/product if they feel they've been tricked into doing something they didn't want to do. Making people feel good about their accomplishments, lowering their stress during wait times – these things build goodwill and trust with your users, making them more likely to come back again and again, resulting in higher satisfaction…er…Net Promoter Scores.

CHAPTER 3 ALL THE FEELS

Cognitive Biases

Simply put, cognitive biases are our tendency, as human beings, to be bad at making objective decisions. At our core, we are wired to make quick decisions based on small amounts of input. For example, if you're walking in the forest and hear rustling bushes, your brain does a quick calculation: rustling leaves = predator that will jump out and eat you. It fires up the adrenaline and before you've had a chance to notice the small bunny hopping out of the bushes, you are halfway down the path, full speed ahead.

And there is a very good reason for this type of fast thinking[4] – survivability. As we were evolving, taking all of the information available into account to make a decision was not a trait that was likely to lead to survival. Running when the sound is caused by a bunny doesn't make sense but running when the sound is a grizzly bear definitely does – and acting quickly on partial information would have saved your life. Today, that fast thinking is less necessary for survival than it once was and has some unintended consequences when it is used.

There was a period of time when the emotional centers of our brain were referred to as our "lizard brain,"[5] which, given the fact that we are creatures of emotion, seems odd, as though somehow these elements made us less human. Emotion is an integral part of how we process information, and that emotion strongly affects our ability to make good decisions. Given all of the stimuli we process in our interactions, it may come as no surprise that we evolved a number of biases and cognitive rules of thumb (called "heuristics") that try to reduce how much we have to process our emotions. We'll talk about biases throughout this book, but for

---

[4] Kahneman, D. (2011). *Thinking, fast and slow.* Farrar, Straus and Giroux.
[5] https://www.amnh.org/exhibitions/brain-the-inside-story/your-emotional-brain/beyond-our-lizard-brain#:~:text=Our%20emotions%20are%20processed%20in,emotions%20and%20make%20complex%20decisions

CHAPTER 3   ALL THE FEELS

now, let's take a look at a few of the more common biases that come into play when designing products (and that you have likely come into contact with as a consumer).

Attentional Bias

Attentional bias refers to our tendency to focus our attention on certain aspects of our environment while ignoring others, prioritizing information that is emotionally salient, personally relevant, or consistent with preexisting beliefs or expectations. One of the earliest experiments used to measure attentional bias is the emotional Stroop test.[6] In one variation of the Stroop test, participants have to name the color of a printed word, and the experimenter measures how long it takes them to name the color. The words are emotionally negative (murder, lies), positive (happy, content), or neutral (chair, table). Because we associate some colors, like red, with some emotions (anger), when there is a mismatch between the color and the emotionality of the word (what cognitive scientists call "valence"), it takes longer for us to process it and name it. This uneven attention leads us to misremember things (ignoring things that weren't as emotionally relevant or were at odds with what we expected or wanted to see) and to undermine our efforts at growth and self-improvement by burying those pieces of information that do not fit our current expectations.

You see this in design when we serve up advertisements that are consistent with items you've purchased or stores you've frequented or in your Facebook feed when you see posts that agree with your point of view. It's important to know how difficult it is to overcome attentional bias to get someone to consider new items or perspectives – the human brain very much defaults to status quo, and it can be very difficult to get someone's attention on something new.

---

[6] Ben-Haim M, Williams P, Howard Z, Mama Y, Eidels A, Algom D. The emotional Stroop task: Assessing cognitive performance under exposure to emotional content. J Vis Exp. 2016;(112):53720. doi:10.3791/53720.

51

CHAPTER 3   ALL THE FEELS

Endowment Effect

The endowment effect is our tendency to place a higher value on objects we own or possess compared to objects we do not. While there are some obvious implications for trade and commerce here (reluctance to part with things you don't use, reluctance to trade for objectively like-valued items), a more interesting implication is the attachment to status quo: That is, we have a preference for maintaining the way things are (viewing our processes and procedures as "possessions") rather than taking risks or making changes, even if those changes could lead to better outcomes overall. One way this bias is leveraged in design is to instill a psychological sense of ownership, which in turn makes you "value" the item or experience more and want to pay more for it.

Using language or audio focused on your vision of yourself (the ubiquitous "See yourself in a <car, vacation spot, clothing>") and imagery that encourages you to picture yourself with the item (an empty seat with an open door, footage shot from the user's POV) leverage the endowment effect. Indeed, the endowment effect also refers to objects that you don't even own. Think about a situation where you miss your flight because you arrived late to the airport. In which instance would you feel worse: if you missed the flight by five minutes or if you missed the flight by five hours. For most people, it is the former because it almost feels like they could have made the flight, whereas in the latter we obviously didn't have a chance. This feeling is brought about by the disappearance of something that we don't have (our seat on the airplane) but could have had (which we deem as more realistic in the five-minutes-too-late but not the five-hours-too-late scenario).

Halo Effect

In the halo effect, a person's overall impression of someone (or something) influences their perceptions of specific attributes or qualities associated with that individual/thing. That is, the halo effect leads us to think that

CHAPTER 3   ALL THE FEELS

because someone has one positive trait, they must have other positive traits as well, without observational evidence. For example, if a student is on time or early to class, the teacher may also think of that student as studious and responsible, even if their grades and performance suggest otherwise. They may, unconsciously, give them consideration when grading or calling on them in class. The halo effect was observed in the 1920s[7] when Edward Thorndike, who was collecting ratings of soldiers by their officers, discovered a correlation between unrelated positive and negative traits – soldiers who were tall and attractive were rated as better, more competent soldiers, regardless of actual performance. Brand reputation relies on this effect – a strong positive brand will provide consumers with the perception that products that they have no experience with at all are great because their impression of the brand is positive.

Observer-Expectancy Effect

The observer-expectancy effect, also known as the experimenter expectancy effect or observer-expectancy bias, refers to the situation in which an experimenter's expectations or beliefs about the outcome of an experiment influence the behavior or responses of the participants. That is, what the researcher believes is going to happen inadvertently shapes the results of the experiment. This can happen when the researchers' expectations about participants' behavior subtly influence their interactions with the participants, leading to unintentional cues or signals that shape the participants' responses or the researchers unconsciously interpret ambiguous responses or data in a way that supports their original assumption.

---

[7] Thorndike, E.L. (1920). "A constant error in psychological ratings". *Journal of Applied Psychology.* **4** (1): 25–29.

CHAPTER 3   ALL THE FEELS

Primacy Effect

The primacy effect is the tendency for people to better remember, and place greater importance on, items or information that are presented earlier in a series, compared to those presented later. That is, when you are exposed to a list of items or a sequence of information, you are more likely to recall and prioritize the items presented at the beginning of the sequence. This effect tends to persist over time, meaning that the impact of early information on memory and decision-making may endure even after exposure to additional information which, in turn, leads to enduring biases or preferences based on the initial presentation of information. This effect is critical to remember when designing lists of items and providing an initial prioritization and is a huge driver in the way social media feeds are constructed. Historically, designers leveraged the area "above the fold" (referring to the area visible on the top of newspapers when they were folded for delivery) for the most critical information that they wanted readers to remember. Today, that translates to placing items on top of a scroll or on the load page for an application.

CHAPTER 3   ALL THE FEELS

Recency Bias

Complementary to the primacy effect, recency bias, also known as the recency effect, is the tendency for people to better remember and give more weight to information or events that occurred most recently, compared to those that occurred earlier. When people are exposed to a series of items or events, they are more likely to recall and prioritize the items or events that occurred last. Both the recency bias and the primacy effect are part of the temporal ordering effect – the tendency for us to remember the beginning and ends of lists better than the middle.

CHAPTER 3   ALL THE FEELS

The Dark Side: Patterns That Are Used to Manipulate

We've all been there. Your "free trial" insists on a credit card number. The X in the upper right-hand corner meant to exit the annoying nag screen actually launches a sign-up process. You check your cart after checking out, and somehow, you've been charged for a service you didn't add. And we all react the same way – screaming thinly veiled obscenities at the screen as if the developers and designers can actually hear us and feel the burning coals we are heaping upon their heads (or maybe that's just us).

WON'T GET FOOLED AGAIN...

As a designer, I find these types of interactions particularly obnoxious because I know that someone somewhere created them deliberately. Make no mistake – dark patterns (designs that take advantage of your cognitive biases to manipulate you or deceive you) are intentional devices, not accidental bad design. And they are everywhere – some to such a degree that we take their presence for granted and even accept that "that's just the way it is." At best, it's annoying. At worst, it's dangerously manipulative, causing people who cannot afford to make purchases to make them unknowingly and costing them money they do not have. But that is starting to change.

In 2022, multiple lawsuits brought by the Federal Trade Commission against companies using dark patterns (such as fees hidden behind required scrolling, falsely routing people to sales sites, and making cancellation of services nearly impossible) resulted in fines of tens of millions of dollars. In March 2023, consumers sued Audible due to free trials that converted to paid subscriptions without notice. In June, the Federal Trade Commission sued Amazon over the onerous process needed to cancel subscription (a pattern often referred to as a "roach motel"

CHAPTER 3   ALL THE FEELS

because of your ability to get in but not get out). In the same month, they settled a lawsuit with Publishers Clearing House resulting in a UI overhaul to address manipulative design patterns.

LUKE: "IS THE DARK SIDE STRONGER?"
YODA: "NO, NO, NO. QUICKER, EASIER, MORE SEDUCTIVE."

57

CHAPTER 3   ALL THE FEELS

Recently, India[8] outlawed the use of 13 dark design patterns, in an update to the consumer protection act, and included language that would make room to outlaw more.

While there's definitely a lot of work to be done in this area, these laws are a start to raising awareness of how design can manipulate people into undesired and harmful actions by exploiting their cognitive biases. Let's take a look at each of the 13 outlawed dark patterns and the cognitive biases that are at work, as well as how to ensure you're avoiding them (preferably at the design system level).

False Urgency

"Only three left! Selling fast!"

"Most Popular Item – limited quantities"

Messages and designs falsely suggesting that a product will soon be gone if the buyer doesn't act quickly plays to action bias (our preference to do something rather than nothing) and scarcity bias (the harder it is to get something, the more we value it). We are afraid we'll miss our opportunity and make purchase decisions impulsively, not reading through the details – basically it's FOMO in your shopping cart. A great example of this is a garden ornament Dr. Baker purchased. It was absolutely adorable – a T-Rex dinosaur wreaking havoc on a bunch of garden gnomes.

"Only one left!"

She hurriedly added it to her cart and checked out...failing to check the size. Her imagined tyrant of the garden arrived at a scant six inches in height. Her broccoli was unimpressed, as was she. If she had taken a

---

[8] https://consumeraffairs.nic.in/sites/default/files/file-uploads/latestnews/The%20Guidelines%20for%20Prevention%20and%20Regulation%20of%20Dark%20Patterns%2C%202023.pdf

CHAPTER 3   ALL THE FEELS

moment to read the dimensions of the tiny terror, she would have realized its diminutive size and passed. But the urgency she felt to purchase it before the seller ran out overcame her usual due diligence.

Avoiding false urgency is fairly straightforward – don't lie about the quantity of something you have or use language to imply a shortage or false importance.

CHAPTER 3   ALL THE FEELS

Basket Sneaking

You complete your checkout and notice that the total is five dollars more than you expected. After looking at the receipt, you see that a donation to Save the Houseplants foundation appears at the end. You don't remember adding it, but if you go back through the workflow, you see there's a check box for the donation that you failed to uncheck. This pattern is known as Basket Sneaking – adding items to the total purchase without the user's consent, including providing a default opt-in for services, donations, and subscription as well as automatic opt-ins for marketing campaigns and emails.

This pattern plays to our attentional bias (our tendency to focus on certain elements and ignore others). By placing the check box at the end of a lengthy checkout, the user is less likely to notice it, and by defaulting to "opt in," that lack of attention causes them to provide permission for something they may not want. A large percentage of the emails in what is likely your overloaded inbox are due to patterns like this – because while you may have noticed the opt-out the first time you purchased something from a website, chances are good you forgot to be on the lookout for it the next time. And now you're suddenly inundated with marketing emails for something you just purchased (seriously, how many copies of "DIY Garden Gnomes" does one person need?).

To avoid basket sneaking, provide optional things like subscriptions and add-ons early on in the checkout process and ensure they are NOT selected by default.

Confirm shaming

"I will stay unsecured."

"I don't care about underprivileged children."

Confirm shaming is using language or visuals to make the user feel shame or fear for not acting or opting out. Surprisingly, the reaction from many designers when discussing design is "but I'm just being funny!"

60

CHAPTER 3   ALL THE FEELS

Unfortunately, humor is very context, culture, and situation-specific and hence extremely tricky to do well and avoid unintended consequences. If you are running a weather app like Carrot, the expectation of snarky humor is established and desired. If you are running a subscription site for coupons...not so much. You have to have a very detailed understanding of your users and consistently establish and reinforce your expectations around communication style and tone.

Social desirability is the underlying bias at work here. Social desirability bias leads us to take action or answer questions in a way that makes us appear more favorable to others. This particular bias is one of the big ones to look out for in survey creation or any kind of self-report research – if an answer to a question is perceived as negative or "bad," respondents are less likely to be truthful or to report on the behavior/ attitude/feeling accurately. Confirm shaming uses this tendency to bully the user into doing the "right" thing.

To avoid confirm shaming, consider your confirmation prompts and remove language that implies wrongdoing on the part of the user. It's important to convey consequences without judgment. "Do you want to delete all of your data?" is good. "Do you want to be a sad, lonely, data-less loser?" not so much.

Forced Action

"To access the results of your quiz, just give us your date of birth, blood type, first born child's name, and email address!"

"To get the full functionality of your purchase, subscribe to our newsletter!"

Forced action is making the user take an action (such as provide additional, unnecessary information, download an app, subscribe to a newsletter) in order to access something they have already purchased or has been promised to them. This pattern leverages the popular sunk-cost bias, which is our tendency to be reluctant to abandon something because

CHAPTER 3   ALL THE FEELS

we have already invested time and/or money into it. Sunk-cost bias leads us to concede to increasingly ridiculous demands for information or action.

Dr. Baker recently took a personality test (who doesn't love a good personality test, am I right?), and at the end of a 15-minute questionnaire, she was asked for her email address so they could send her the results along with marketing to buy the service and books to tell her how to interpret those results. At no time before or during the lengthy Q&A was she told that she would need to pay anything for my results or provide her email. She declined to give them her information, but that meant she also didn't get the results of her labor filling out the test.

To avoid forced action patterns, ensure any requests for information or money are optional or are well-documented upfront. If the lamp doesn't work without also purchasing a cord separately, make sure that's known – and never require an email after purchase.

Subscription Trap

"To unsubscribe, simply contact our support team between the hours of 3:00am–3:05am local time, and provide the account number printed on your original receipt in 1-point type on the third page. Good luck!"

"To sign up for your free trial, just give us a valid credit card number so we can start charging your subscription the moment the trial is over but without asking you if you actually want to continue."

Subscription traps make it ridiculously hard for someone to cancel a subscription by hiding information on how to cancel or requiring payment information for free trials that will automatically become subscriptions, forcing users to remember when their trial is up in order to cancel it. Subscription traps rely less on cognitive biases and more on our forgetfulness and being easily frustrated. By putting the burden of cancellation on the user and making that cancellation as difficult as possible, companies are hoping for that "one more month" they can

CHAPTER 3   ALL THE FEELS

squeeze out of consumers – something we've all experienced when March rolls around and you haven't used that gym membership you got in January for the last two months (because, frankly, it was kind of cold in the mornings and dark and your bed was super comfy, plus you stayed up too late working on that project, and didn't you read somewhere that sleep was super important?), but you also can't remember exactly how to cancel it – but you're pretty sure it involved breaking a blood oath or something similar. Requiring a significant amount of notice in writing (not email), coming into the business "in person," needing an original receipt, and so on – all of these lead to a difficult experience in cancellation. Unfortunately, designing experiences in this way is itself leaning into a cognitive bias called present bias. Present bias is the tendency for people (and companies!) to value short-term gain at the expense of long-term well-being (this could also apply to those excuses around not using that gym membership).

Creating such a difficult and negative experience for subscription cancellations violates Daniel Kahneman's well-researched Peak-End Rule.[9] The Peak-End Rule states that our impression of an experience is dictated by the highest or lowest part of that experience and the end. In other words, how someone thinks about your product is not based on an overall ebb and flow of good and bad experiences – it's based on the best/ worst thing that happens averaged with the last thing that happens. Not the first – the last. Failing to design for endings is a huge gap in design and something worth thinking about when you want to make it hard for your competitors to keep the customers they may have enticed away from you or get good reviews from existing and past customers.

To avoid subscription traps, make unsubscribing simple and easy to find and never require financial commitments (even future ones) for a free trial. Ensure the burden is on your program/experience, not on the user.

---

[9] https://link.springer.com/content/pdf/10.3758/PBR.15.1.96.pdf

CHAPTER 3   ALL THE FEELS

Interface Interference

You get an unwanted ad pop-up and promptly click the x in the upper right-hand corner. Instead of closing the dialog as you expected, that click launches a new tab, replete with blink tags, loud music, and flashy graphics for a product you've never heard of. Because it's a new tab, there's not a back button to escape the visual and auditory cacophony assaulting your senses. Interface interference is obscuring/highlighting interface items or taking common interface items and making them behave in ways counter to what is expected in order to misdirect the user's actions. This pattern takes advantage of our expectations to trick us into doing something we didn't mean to or shouldn't do. A close button that is the same color as the background, rendering it invisible and making it appear that the only option is the call-to-action "buy," tries to trick us into downloading or purchasing something we didn't want. A close dialog button that actually launches a new tab (or several tabs) with no obvious way out tries to get us to stay on the site and hopefully be tempted by more advertisements. In the following example from 2016, clicking the red X in the upper right-hand corner is the equivalent of saying "yes I want to upgrade."

CHAPTER 3   ALL THE FEELS

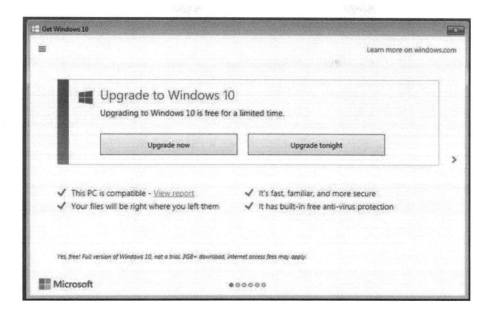

All of these patterns are intended to make you take an action you did not want or intend to take, and that will benefit the program, not you. Interface interference relies on our ever-present information overload by violating conventions we rely on to automate our actions. As a result, we become more susceptible to any number of cognitive biases which can then lead us to make poor decisions.

To avoid interface interference, you can follow established conventions for interface interactions and adhere to the well-documented usability heuristics[10] like those from Jakob Nielsen.

Bait and Switch

"The flower arrangement you selected is not available in your area – try this one that's $20 more instead!"

---
[10] https://www.nngroup.com/articles/ten-usability-heuristics/

CHAPTER 3   ALL THE FEELS

"You've got great taste! Unfortunately, all of the slots for Attracting Gnomes to Your Garden 101 are taken. We recommend taking the entire series of Gnome Care and Feeding instead for only seven additional payments for $49.99 each!"

Bait and switch patterns suggest that you can get a particular outcome, but when it comes time to commit, an alternate, more expensive, outcome is offered. This pattern operates off of a couple of cognitive biases working together. Loss aversion bias is our tendency to want to avoid loss rather than gain something. You've decided you want an item – when it's shown as unavailable, you automatically want to avoid that loss. FOMO (Fear of Missing Out) also comes into play – we have a very real aversion to missing out on something that everyone else is in on. When you find something in your cart is gone because it's so popular, you're more likely to get the substitute to avoid being "left behind."

More subtle variations of this dark pattern include listing products that don't yet exist but that you intend to implement. "Available in future releases" "Coming soon!" – without a specific time period associated with "soon" or "future," this type of language falls into bait and switch by luring users into purchasing a product on the promise that they will be getting something in the next couple of weeks when, in fact, the real time line is several months or years. Travel sites often employee this design – in the following example, the advertised special is quite a bit different than the actual price when booking.

CHAPTER 3   ALL THE FEELS

To avoid bait and switch patterns, ensure you are accurately representing the availability of items and if you are truly out between the time they "add to cart" and check out, do not automatically add items to the basket as substitutes. Instead, ask if they would like to see similar offerings or show a selection of similarly priced items (not in the basket) that could be a good substitute. If you have an item or feature that you want users to know is being considered for a future release, ensure your language is specific and accurate (will release in February 2023). Save "under-consideration" communications for existing customers, and be clear that they are not yet a commitment.

Drip Pricing

"Only 99.99 – except there's also 59.99 fee for parking and 199.99 for walking through the lobby."

Adding fees or insisting on additional money for something after you have entered the checkout process or advertising only a portion of what you will have to pay for a given product or service instead of providing full disclosure on the price is called drip pricing. Drip pricing is a long-standing strategy in many industries (most commonly the hospitality industry) to reel in customers with a price they find attractive but then reveal, after they are at or near completion of the transaction, fees and costs that make the final price much higher than the advertised original price. This pattern leverages your present/immediacy bias – our learning to focus on the current situation rather than focus on future impact. That is, once you are in the middle of a complicated checkout procedure and find out about the extra cost, going back and doing it all over again feels like more work than the extra fees are worth. It also leverages your loss aversion bias – our tendency to find losing something more painful than gaining the same thing is pleasant. Think about having chosen first-class tickets because they were only $10 more than coach, only to find out there

67

CHAPTER 3   ALL THE FEELS

are $100 in fees associated with first class right before you book – the potential loss of the first-class tickets feels worse than the gain of more affordable seating that is in your budget. It's worth noting that California has outlawed (starting July 1, 2024) the use of drip pricing[11] and the US Federal government is considering new regulations around this tactic (also referred to as "junk fees").

Shopping Cart (1 item)

Clear All

Subtotal*

$273.60

(*Excluding taxes)

Summary

1/21 room		$254.00	$228.60
Room Estimated Tax			+$30.59
Resort Fee ⓘ			+$45.00
Resort Fee Estimated Tax			+$6.02
Room Subtotal			**$310.21**

To avoid this type of pattern, advertise the actual, real price including all fees, up front. Never hide necessary fees or costs until later in the checkout process.

---

[11] https://www.jenner.com/en/news-insights/publications/new-california-law-targets-drip-pricing

CHAPTER 3   ALL THE FEELS

Disguised Advertisements

You find yourself on a site offering a free version of some software you've been dying to try. You click download only to be taken to another site entirely (plus no software!). Later you are on a news site, scrolling through the feed. Formatted exactly like the actual news articles is an ad for discounted MacBooks. These are examples of disguised advertisements which is a form of interface interference. And like interface interference, it relies on our reliance on mental shortcuts (in this case common UI patterns) to help us handle information overload and the representative heuristic (our desire for there to be a pattern where one may or may not exist). Disguised advertisements usually appear in crowded, information-rich, pages and can be as subtle as a link in the middle of the text that, instead of leading to an explanation, takes you to a product page. These types of advertisements are addressed specifically by the FTC[12] and can provide a serious liability risk for those using them. In the following image, this now-defunct website offers two download buttons – one of which launches advertisements.

---

[12] https://www.ftc.gov/system/files/documents/public_statements/896923/151222deceptiveenforcement.pdf

CHAPTER 3   ALL THE FEELS

To avoid using disguised advertising, always label ads as such. Ideally, create a visual distinction that makes it clear that the content is an advertisement and provide corresponding ALT tags to ensure that even if the content is not being consumed visually, it's clear that this piece is an ad.

Nagging

Nagging is the practice of pestering, disrupting, or annoying users to make a commercial gain. Patterns that continually ask to enable notifications or provide information that attempt to increase the time and money you spend in an application are considered nagging. Part of a larger architecture known as nudges, this pattern is intended to guide or move the user in a particular direction. The psychology behind nudges

CHAPTER 3 ALL THE FEELS

(and nagging) was made popular in the book *Nudge: Improving Decisions about Health, Wealth, and Happiness* by Thaler and Sunstein and applies to many disciplines. One of the most famous examples of nudging occurred in airport urinals in Amsterdam. The airport administration had realistic pictures of houseflies painted near the drain of each urinal (intended to improve aim and lower cleaning costs; yes, it worked).

BUT WAIT, THERE'S MORE...

A friend of mine, let's call her "Ethel," really enjoys video calling but is not a fan of Facetime (long story). Instead Ethel uses a "free" communication app that is video-only. Ethel asked me to install it so she could talk to me via the app, so I did. After a perfectly annoying installation process (do you really need to know my address?), I managed to get it installed. It promptly, and without asking, defaulted to enabling badging notifications. While normally I wouldn't mind this from a communications application, this one insists on badging (the small red dot in the upper right-hand corner of the app on your phone screen) to indicate that I haven't explored the entire application, including paying for an upgrade. Nothing I can do gets rid of it, short of disabling all notifications from the app (which I did).

Other examples include putting the choices the grocery store wants you to choose at eye level, providing a default selection in checkout options (like an option to subscribe to the newsletter), and using AI to show ads based on past browsing behavior. Nagging relies on moving your decision-making from the reflective side of your decision system to the automatic side of your decision system. Automatic decisions are easy, and we like them because we don't have to think about them. Reflective decisions are harder, because we have to consider and be more intentional

71

CHAPTER 3   ALL THE FEELS

in our thinking. (see Kahneman's *Thinking Fast and Slow* for an in-depth look at this). When we are making automatic decisions, we are much more susceptible to cognitive biases like anchoring (taking the first option presented as the baseline and comparing all options afterward to that). Nagging, in particular, is intended to guide us into taking action by relying on our already overtasked brain's information overload to be willing to pay to just make it go away.

It's a strategy that Dr. Baker often uses on her kids to get them to clean their rooms or eat their veggies (and that they use on her to get ice cream). And while nagging to get your teenager to pick up their stinky clothes off the floor may be marginally effective, widely accepted, and ok, nagging to get your customers to upgrade to a paid version or add a new feature is not ok, regardless of its effectiveness or acceptance.

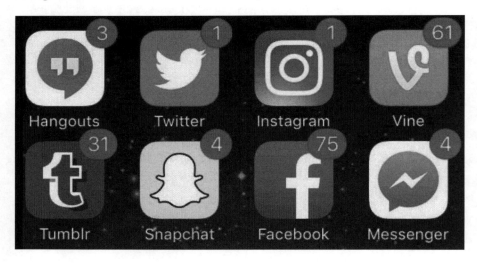

To avoid nagging, ensure that your reminders, nudges, badging, messaging, and so on are off by default and it is clear when it is turned on what type of messaging is going to be provided (marketing, system needs, etc.).

CHAPTER 3   ALL THE FEELS

Trick Questions

I don't know half of you half as well as I should like; and I like less than half of you half as well as you deserve.

—Bilbo Baggins, The Fellowship of the Ring

"To opt out of emails from us, please do not select any of the following options:" People only read about 20% of the text on a given web page.[13] Trick questions take advantage of this lack of attention (known as inattentional blindness) by using our daily information overload to make us think we are agreeing to one thing when, in fact, we are agreeing to the opposite. Inattentional blindness occurs when we are focused on one thing and are unable to process unexpected events. A fun example of this is the invisible gorilla experiment[14] wherein participants are asked to watch a video of two teams passing a basketball back and forth and count the number of times the ball switches teams. In the middle of the video, a person in a gorilla suit calmly walks through the game, pauses, beats their chest, then exits. Afterward, about half of the participants did not recall seeing the gorilla at all. Trick questions use this bias by presenting language or selections in a familiar way but reversing the meaning. Using double negatives or convoluted language and presenting choices that are the opposite of the expected (select your email address to opt out of emails) are all examples of trick questions. To avoid trick questions, ensure you are using plain language that clearly states the options and consequences your users will have from selecting those options at a sixth grade reading level or lower. Use standard, expected design patterns to support decision-making without relying entirely on text. Remember, your

---

[13] https://www.nngroup.com/articles/how-little-do-users-read/
[14] http://www.theinvisiblegorilla.com/gorilla_experiment.html

CHAPTER 3   ALL THE FEELS

audience may have your primary language as a second language, may be neurodiverse, or may have competing demands on their attention when dealing with your form (like THAT ever happens...).

SAAS Billing

The advent of SAAS (Software as a Service) brought with it the idea of "renting" software services – that is, having a recurring charge/subscription that affords you access to a feature or set of features rather than buying, downloading, and owning an instance of the software outright. This practice has brought a lot of flexibility and accessibility for both users and creators, paving the way for frequent updates and feature additions that do not require lengthy downloads and potential rewrites of custom configurations as well as lower prices and commitment to a given platform or service. Unfortunately, it also brought the practice of auto renewal and billing. With auto renewal, you are automatically charged at a given interval (monthly, yearly, etc.) for a service, under the assumption that you wish to continue using that service. If you do, in fact, want to continue using it, that's great! However, if you do not and if you have, perhaps, forgotten that you signed up for the service in the first place, it can be annoying and upsetting. These services often do not alert the user each time they get billed, relying on the forgetfulness of those users to get more money. Unwanted recurring charges are so common now that some banking software programs have added features to detect and review recurring charges to help users control the potential drain. Being proactive and alerting users of upcoming charges and how to cancel is an easy way to avoid this dark pattern.

Rogue Malware

Malware – programs that invade your computer to steal or delete your data – are the new bogeymen of our time. The very real threat they present can be felt at the country, company, or individual level, and the damage they do is devastating. In January 2024, Russian hackers stole 2.5 million documents from the Australian government.[15] Malware has been used by both Russia and the Ukraine to attack infrastructure in the opposing country during the war. Companies have spent millions in dealing with the fallout of these types of attacks, in money, time, and loss of reputation. Individuals are also often targeted – having their data, identity, and/or money stolen. The 2024 movie, *The Beekeeper*, deals directly with this type of threat (albeit in a satisfyingly, if unrealistically, kick-ass way) when the main character's friend is subjected to a malware attack and the main character proceeds to go on a rampage against the shadow corporation responsible. While not an accurate depiction of malware attacks (and the response to them), this popular movie demonstrates how much this threat has become a very real part of our everyday lives. When it comes to malware and dark patterns, we are talking about a very specific type of malware called rogue malware. Rogue malware tricks users into thinking their computer has become infected with a virus and then charges them to remove the nonexistent virus. This pattern does not rely on a cognitive bias per se (although the patterns used in the malware may use biases such as the illusory truth effect in which something is perceived as true if it has been stated multiple times or is easy to process) but rather relies on our fear of malware and the loss of data that could be incurred.

The Dark Side and AI

Dark patterns are everywhere. They are used to manipulate our behavior to the advantage of someone else, usually a company looking to get us to buy more of their product or service. As users, it's important that we understand these patterns so that we can make informed decisions about what we want to do and not allow ourselves to be manipulated.

---

[15] https://www.csis.org/programs/strategic-technologies-program/ significant-cyber-incidents

CHAPTER 3 ALL THE FEELS

As designers, it's incumbent on us to ensure we understand why our designs work on a cognitive level, as well as the context in which our designs are being used. In this way, we can be both more effective in our design work and thoughtful, ensuring that our designs do not go to the dark side. Yoda was right – the dark side is quicker, easier, and more seductive because it will get you fast results. But it is not stronger – dark patterns are quick fixes with immediate returns, but they will anger your users and discourage them from coming back.

This is especially important when we start to think about using generative AI to accelerate design work. When we think about how generative AI determines its responses, we must remember that while it has a vast sea of data to draw from, ultimately it lacks context and the judgment that accompanies that context. Dark patterns are manipulative but they are also very very effective. AI-generated designs are likely to suggest dark patterns – perhaps even favor them – because it lacks the context of a human to determine WHY the patterns are effective and whether that is a good approach in the long term. Asking AI to create a consumer funnel that optimizes conversions is likely to result in a funnel with no exits. Effective? Yes. Good design that encourages repeat customers? No. This issue underlines the importance of understanding WHY certain design patterns are effective so that the designers can make good, informed decisions on when, where, and how to use them.

Recap

- Emotions are an important evolutionary adaptation that help us survive.

- Cognitive processes – from decision-making to memory – all are the result of a complex interplay between our "fast" regions (the emotional ones) and our "slow" ones (those that involve more processing).

CHAPTER 3   ALL THE FEELS

- Our brains have developed ways to deal with a lot of information quickly that involves both emotions and cognition – but this comes at a cost.

- The tricks our brains play on us are called cognitive biases.

- Understanding common cognitive biases can inform better and more effective design practices.

- Dark patterns are designs that exploit cognitive biases to manipulate people.

- Because humans designed the vast library of information that AI uses for results, you need to know that it may (and likely will) introduce these biases into design

Before You Go…

At the start of this chapter, we referred to the fact that once upon a time some of our basic responses and urges, like emotions, were referred to as being part of our "lizard brain." This idea was first put forth by the neuroanatomist Paul MacLean who put forward his "triune" (i.e., three-part) brain theory after noting that reptiles and mammals share common neurological structures.[16] Though not false *per se*, it is important to note that though a crocodile and a human do both have structures in common like the cerebellum, hippocampus, and others; the detailed and complex emotions exhibited by only a few species, including humans, are the result of millions of years of differential evolution (i.e., meaning

---

[16] https://www.biotechniques.com/neuroscience/10x_sptl_so_no-we-dont-have-a-lizard-brain/#:~:text=In%20the%201960s%2C%20Paul%20MacLean,after%20life%20moved%20to%20land

CHAPTER 3 ALL THE FEELS

the time since the two species diverged from a common ancestor). To put this into context, though we often think of rats and mice as being quite similar, they actually became two quite distinct species over 20 million years ago, an amount of time that results in robust, significant differences in behavior and the underlying structure and neurochemistry of their brains[17] (MacLean wasn't able to make comparisons at the level of specificity that modern neuroanatomists are able to employ). Given that mammals diverged from reptiles over 200 million years ago and that our evolution more closely followed that of other warm-blooded species, it would have been more accurate to compare us with avian species, with whom we share more behavioral characteristics[18]...but then, we would be talking about our actions being the result of our "bird brains," and, though comical, nobody wants that.

---

[17] Rekart, J. L., Sandoval, C. J., & Routtenberg, A. (2007). Learning-induced axonal remodeling: evolutionary divergence and conservation of two components of the mossy fiber system within Rodentia. *Neurobiology of learning and memory, 87*(2), 225–235.

[18] Tomasello, M. (2022). *The evolution of agency: Behavioral organization from lizards to humans.* MIT Press.

CHAPTER 4

Being Part of Something

RB: Belongingness is a weird word. Why must we always come up with new labels for things? Feels sus.

JLR: Agreed. But the need to feel like you belong to something, regardless of what we're now calling it, could be described as primal. It is instinctual, and as we'll explore shortly, being part of a group is not only a uniquely human trait but an important one.

RB: True. And as such, it's critical to understand when it comes to designing for humans. Our desire – or should I say need – to belong is an important motivator for many of the things we do. For good or for bad.

JLR: Like feeling the need to identify as a Swiftie?

RB: Just like that.

What Does It Mean to Belong? Evolutionary Foundations of Human Groups

As mentioned in Chapter 3, humans lack many of the defensive and physical adaptations that enable the survival of other animals. What we do have, however, are complex social structures that facilitate the ability of groups to not only survive but thrive. Though we know that complicated social structures are also found among our primate cousins, like

© Jerome L. Rekart and Rebecca Baker 2025
J. L. Rekart and R. Baker, *Designing for Human Intelligence in an Artificial Intelligence World*,
https://doi.org/10.1007/979-8-8688-1418-1_4

CHAPTER 4 BEING PART OF SOMETHING

chimpanzees and bonobos, the sheer size of human intimate networks, meaning those with which people have regular, sustained, and meaningful interactions (i.e., where we learn things about one another and use those in discourse), easily dwarf those of our nonhuman kin. The number of such relations, which was extrapolated by the British Anthropologist, Robin Dunbar (for whom it is named; i.e., "Dunbar's number"), maxes out at around 150 individuals.[1] Dunbar has theorized that the size of such groups may have been a selective pressure on early hominins (i.e., human ancestors, like *Homo erectus*), which caused the increases in neocortical size believed to be responsible for language and other forms of intelligent cognition that we all enjoy today.

Though the debate about whether the size of early human groups led to our use of language or whether our development of language facilitated the development of larger communities has yet to be resolved, what is certain is that the development of advanced forms of language – both oral and written – made it possible for humans to interact with numbers of individuals far greater than those observed in any other species on the planet.

It is important here to clarify that by "interact" we mean that two individuals in some way share an experience with some form of explicit or implicit acknowledgement. For example, any of the following would all constitute a human interaction of this sort:

- A phone call between a customer service representative and a customer (whether irate or not...because unfortunately, the majority of content customers just don't call)

- A nervous smile you flash at someone across the aisle of an airplane when you hit a bump of turbulence

---

[1] Dunbar, R. I. M. (1993). Coevolution of neocortical size, group size and language in humans. *Behavioral and Brain Sciences, 16*(4), 681–735.

CHAPTER 4   BEING PART OF SOMETHING

- A nod of your head at something a tour guide says while on a walking tour of a university you are considering attending

This notion of interaction is critical. Certainly, insect species exist in communities that exceed Dunbar's number, yet most are not considered to have advanced intelligence or complex language on par with humans. For example, in the summer of 2024, two separate cicada broods (XIII and XVII) emerged from the ground after 13- and 17-year slumbers, respectively, and though they numbered in the trillions,[2] to the best of our knowledge, they were not engaged in interactions the way we have defined them.[3] Although, to be fair, if they were glancing nervously at each other...would we know?

Nor do they belong to groups other than those defined by their own brood (i.e., XIII) or as dictated by geography. Humans, on the other hand, belong to more groups than any other species on this planet.[4] Think about all of the ways that you define yourself and how many of those to which you belong are meaningful.

For example, if we identify a few ways that we see ourselves, it may look like this:

Dr. Baker: I "belong" to a lot of different groups, and I use that to relate to other people who also identify with those groups:

- A mother

- A writer

- A dungeon master

---

[2] Bosman, J. (2024, May 4). Here Come a Trillion Cicadas. The Midwest is Abuzz. *The New York Times.* https://www.nytimes.com/2024/05/04/us/cicada-emergece-illinois-midwest.html.

[3] Although, to be fair, if they were glancing nervously at each other...would we know?

[4] Moffett, M. W. (2013). Human identity and the evolution of societies. *Human Nature, 24,* 219–267.

CHAPTER 4   BEING PART OF SOMETHING

- A fossil hunter
- A Texan
- A scientist

Dr. Rekart: I see myself as having an identity that suggests membership in both formal and informal groups. I am

- A father
- A citizen of the United States of America
- A Midwesterner by birth, a New Englander by choice
- A Cubs fan
- A Deadhead
- An alumni of Indiana University, Northwestern University, and MIT

These groups are defined by geography, past experiences, things that are liked or enjoyed, and professional associations. And they are not not even close to exhausting the sheer number of groups, conclaves, or tribes (or gangs or posses or...) to which either of us feels a kinship. We could try to understand the similarities between those groups by aligning them along categories like core identity (e.g., mother, father, scientist), geography (e.g., Texas, the Midwest of the United States), and affiliations and interests (e.g., Cubs fan, dungeon master...although either of these could also be subsumed under "core"), but ultimately, the key is that we consider each to be a part of what represents us and what we represent. More than just a one-way street (e.g., "I get 'X' from this relationship, group, or community"), the key to social groupings and interactions is mutuality and reciprocity[5] (e.g., "I get 'X' from this relationship, group,

---

[5] Molm, L. D., Collett, J. L., & Schaefer, D. R. (2007). Building solidarity through generalized exchange: A theory of reciprocity. *American journal of sociology*, *113*(1), 205–242.

CHAPTER 4   BEING PART OF SOMETHING

or community and am able to contribute 'Y'"). Thus, belonging requires not only that we know ourselves but that we're able to recognize and understand what others want, desire, and are feeling, as well.

Mirror Neurons and the Art of Reflection

Have you ever wondered why you get the urge to yawn after you see someone else yawning? This phenomenon, which is referred to as contagious yawning (for real), is caused in part by a specialized set of neurons in our motor system (i.e., the part of the brain responsible for movement) that actually get activated when we see someone else perform an action.[6] These mimicry cells, which are also associated with parts of the brain that underlie emotion and sensation, are referred to as "mirror neurons."

---

[6] Haker, H., Kawohl, W., Herwig, U., & Rössler, W. (2013). Mirror neuron activity during contagious yawning—an fMRI study. *Brain imaging and behavior, 7*, 28-34.

83

CHAPTER 4   BEING PART OF SOMETHING

These neurons are, in a way, the building blocks of our social nature. Some evidence suggests that these specialized neurons are not only why yawns are contagious but also why we empathize with others.[7] This is why (in part) we experience joy when our team wins and grief when characters on television die[8] (poor, poor Hodor). Put simply, this is why stories are such powerful tools for evoking belonging and empathy.

Indeed, the power of stories is so strong that studies have shown that feelings of empathy toward individuals of what are called "out-groups" (e.g., for white students, this would be students of color; for Muslim students, it might be those who are Jewish) are increased (i.e., more acceptance and tolerance) through storytelling that describes the experiences of members of those groups.[9]

Though we will dedicate much more time to the power of storytelling to increase engagement, facilitate remembering, and engender empathy in Chapter 8, it is important to note here that storytelling is one of the ways that we use our language, which is quite advanced, to reach mirror neurons (a basic cell type in some mammals and birds). This, in turn, helps us understand, from a neurological perspective, the basis of belonging. And these mirror neurons – and the empathy they engender – are not a feature of AI.

The Psychology of Belonging

Certainly in today's social media-saturated world, though the need to be "liked" or "followed" could be summarily dismissed as just a generational trend that will someday suffer the same fate as bell bottoms (ugh), Trapper

---

[7] Bonini, L., Rotunno, C., Arcuri, E., & Gallese, V. (2022). Mirror neurons 30 years later: implications and applications. *Trends in cognitive sciences*, *26*(9), 767–781.

[8] Singer, T., & Lamm, C. (2009). The social neuroscience of empathy. *Annals of the new York Academy of Sciences*, *1156*(1), 81–96.

[9] Weisz, E., & Zaki, J. (2017). Empathy building interventions: A review of existing work and suggestions for future directions. *The Oxford handbook of compassion science*, 205–217.

CHAPTER 4   BEING PART OF SOMETHING

Keepers (fun and functional), and frosted tips (who thought this was a good idea?), the underlying desire to be seen and accepted will not. Indeed, the desire to belong is believed to be a foundational motivator of human behavior.[10]

For our purposes, "belonging" can be thought of as the feeling one gets when they believe they are accepted. In other words, the validation that occurs when someone feels like they have value and that they are welcomed – flaws and all – into a group.[11]

Belongingness can be thought of in two different ways: as either a state or a trait. Given difficulties in measuring the latter (i.e., like a character or personality trait), we will primarily focus here on the former, which is to say a short-term, transient feeling that one has about whether they belong to a group, place, or community in a given moment rather than a stable, long-term disposition.[12] For example, you may feel like an outsider or feel alone (the opposite of belonging) when first arriving at a professional conference (even though your career trajectory and professional interests are similar to those of other attendees). However, all of those feelings could quickly subside and be supplanted by a feeling of belonging when you recognize a colleague from another institution, even if your only interaction is the exchange of smiles with one another from across a crowded convention hall.

---

[10] Allen, K. A., Gray, D. L., Baumeister, R. F., & Leary, M. R. (2022). The need to belong: A deep dive into the origins, implications, and future of a foundational construct. *Educational psychology review, 34*(2), 1133–1156.

[11] Lambert, N. M., Stillman, T. F., Hicks, J. A., Kamble, S., Baumeister, R. F., & Fincham, F. D. (2013). To belong is to matter: Sense of belonging enhances meaning in life. *Personality and social psychology bulletin, 39*(11), 1418–1427.

[12] Allen, K. A., Kern, M. L., Rozek, C. S., McInerney, D. M., & Slavich, G. M. (2021). Belonging: A review of conceptual issues, an integrative framework, and directions for future research. *Australian journal of psychology, 73*(1), 87–102.

CHAPTER 4   BEING PART OF SOMETHING

Though this example provides one instance of how belonging can be cultivated in a given moment, it does not exhaustively cover all of the ways to instill this feeling. Researchers Hirsch and Clark[13] have identified four "pathways" to achieving belonging:

- **Minor Sociability**: Boost in a sense of belonging through minor or incidental social interactions with others, as was described in the example (above) of having one's sense of belonging increased at a convention through a brief, distant exchange of smiles.

- **Communal Relationships**: A long-term path that requires mutual trust and respect built up over time and repeated encounters that justify membership in this group (even if only a dyad or partnership; e.g., feeling like one "belongs" to a married partnership).

- **Group Membership**: The easiest means to achieve belonging, as it can occur through involuntary means (e.g., being a heterosexual man) or voluntary (e.g., being a Cubs fan).

- **General Approbation**: This form of belonging is gained through accomplishment and/or achievement, as well as through associations with individuals or groups that are esteemed. This explains why devout consumers of some brands, like Apple, are so proud to let you know that they are users of their products.

From a design perspective, at least three of these paths should be considered if building a sense of belonging into your designs is desired. The fourth, and arguably most difficult, is that of communal relationships.

---

[13] Hirsch, J. L., & Clark, M. S. (2019). Multiple paths to belonging that we should study together. *Perspectives on Psychological Science, 14*(2), 238–255.

CHAPTER 4   BEING PART OF SOMETHING

Though a platform may facilitate communal relationships, and therefore a sense of belonging, it doesn't mean that users feel that sense for the platform itself. For example, on Reddit, many users feel a sense of camaraderie with others in a subreddit or group, but that doesn't translate to brand loyalty or mean that the users wouldn't be just as comfortable and feel as much belonging if they were to interact with others using other platforms or sites.

Thus, given the three remaining paths, let's look at some ways to design experiences to facilitate belonging.

Let's Play the Belonging Game

So you might be wondering why we have stuff about games in a chapter about belonging. Good question! It turns out that games provide a way for people from different places, social strata, and ages to come together and gain connection. Most games are inherently social, designed to be played with others, either in person or online. The interactions within the games within a psychologically safe environment can lead to the formation of new relationships by providing the players with a shared experience.

Power of Play

Play is a powerful tool that helps us connect with others on multiple levels – social, emotional, and cognitive – because it involves shared experiences, communication, and collaboration. Games are a construct that allow us to play and experience these connections.

Shared Experiences and Joy

What's the big deal about shared experiences? Think about the last time you took the train or attended a sporting event. During each of these activities, you are surrounded by strangers who are going through the same boredom, annoyance, or excitement as you are. For example:

87

CHAPTER 4   BEING PART OF SOMETHING

- As you sit on the train, it rolls to an unexpected stop – the announcer says there will be a slight delay to clear something on the track ahead. You momentarily share a glance and an eye roll with one of your fellow passengers.

- Your team scores a goal in the last seconds of the game – in your mutual excitement, you embrace the fan sitting next to you as you scream your support.

Shared experiences can provide a foundation for connection. Games set up an environment in which shared experiences are the norm – the cooperative striving toward a common goal in a role playing game, a friendly competitive game of dominoes or cards, or the shared laughter of a party game – these experiences provide a shared set of stories, jokes, and memories with other people. And they do so in a regulated environment, safe because the rules of engagement aren't simply known, they're documented. This, in turn, supports connecting people from different walks of life (different cultures, different backgrounds) in an equalizing and inclusive framework.

Unsurprisingly, games often lend themselves to community building. Players who have a passion for a particular game (e.g., *Dungeons & Dragons* or pickleball) seek out others who are also interested in these games as a common ground. These communities let players discuss strategies, share stories, and support each other.

Communication and Interaction

Play requires participants to communicate, whether through spoken language, body language, or even unspoken understanding. This interaction helps people to understand each other better and build rapport. Many games and activities require players to negotiate rules, make decisions together, or cooperate to achieve a common goal. This fosters teamwork and

88

CHAPTER 4 BEING PART OF SOMETHING

helps people learn to work together effectively. Collaborating to solve these problems helps people connect as they share the satisfaction of overcoming obstacles. The sense of accomplishment when a group succeeds in a challenging game or activity can strengthen bonds. Shared victories create lasting memories and reinforce group identity.

Every Sunday night in Dr. Baker's house is game night. There, her family plays everything from *Exploding Kittens* to *Dungeons & Dragons* to *Scrabble* to *Catopoly* (admittedly her least favorite – the kittybox schtick is just a step too far). The entire family looks forward to these games as it is an opportunity for them to connect outside of the "get your chores done", "have you finished your homework?," "what's for dinner?" day-to-day of family life. It's a set of interactions geared toward having fun together. The communication fostered in these weekly games is wonderfully connecting, because it is regulated by the game not by their normal routine.

Similarly, Dr. Baker has had good luck creating community at work using games. In an entirely remote setting, she set up a weekly "Mayhem and Mingling" session. For 30 minutes at lunch each Friday, her team dialed into a call and played games together like *Drawsaurus* (a free online Pictionary-type game). Attendance was optional, but the managers always made an effort to show up, and most of the team did as well. People who were strangers to one another came together, laughing, over some beautifully awful artwork, and built a common language through shared experiences. Her personal favorite was the "banana for scale" – when someone was struggling to draw anything, they would include a drawing of a banana and declare "banana for scale" so you could figure out how big the thing actually was. This sense of camaraderie engendered the belonging that eventually led the group to continue to be cohesive even after a layoff.

CHAPTER 4   BEING PART OF SOMETHING

Building Trust and Empathy Across Cultures

In cooperative play, especially in team-based games, players must rely on one another to succeed. This reliance builds trust as they learn to depend on each other's strengths and skills. Playing with others allows individuals to see things from different perspectives, especially in role-playing games or scenarios where players must put themselves in someone else's shoes.

> **PRETEND TO BE SOMEONE ELSE AND CALL ME IN THE AM**
>
> Role-play has also been shown to increase the empathy of healthcare practitioners after they've spent time in the role of patients.[14] The efficacy of findings like these show much promise for other emotionally draining jobs, such as those in customer service[15].

---

[14] Bearman, M., Palermo, C., Allen, L. M., & Williams, B. (2015). Learning empathy through simulation: a systematic literature review. Simulation in healthcare, 10(5), 308–319.

[15] Groth, M., Wu, Y., Nguyen, H., & Johnson, A. (2019). The moment of truth: A review, synthesis, and research agenda for the customer service experience. Annual Review of Organizational Psychology and Organizational Behavior, 6(1), 89–113.

CHAPTER 4   BEING PART OF SOMETHING

One of us (Dr. Baker) loves *Dungeons & Dragons*, one of the most popular role-playing games on the market (fun fact for fellow players: she was actually the editor on the *Vilhon Reach* module). When she introduces a new player to the game, she runs them through an imaginary scenario in which they are a character in a dark forest, riding a horse. She sets the scene by telling them:

"You hear a noise in the bushes – what do you do?"

What they choose varies a lot depending on their personality and culture. Some spur their horse to run away, some draw their swords to be ready for what happens next, some dismount from their horse to protect it, and some charge into the bushes. The acceptance of their action (there is no wrong answer) helps build trust that provides a sense of freedom and belonging in the setting even though they may have never played before. Similar to the empowerment we get when we get a "like" in social media, the acceptance that games promote help move past cultural or social differences. Interestingly, this same type of acceptance is taught in improv comedy. Whatever the other person you are working with says, you accept it and work from there (i.e., "Yes, and..."). That acceptance is what makes the skit work – and makes the audience laugh.

If someone came up to you and asked what you are doing with a cat on your head, your response isn't amusing if you claim you don't have a cat on your head. By the rules of the improv game, you have to accept that there is a cat on your head and exclaim "A cat?! I thought my hair was just being excessively cheeky this morning! No wonder I had so much trouble washing it." Collaborative play like this often sparks creativity and innovation, as players build on each other's ideas. This collaborative creativity fosters a deeper connection as participants feel valued and understood.

CHAPTER 4 BEING PART OF SOMETHING

Gamify All the Things (or Not)

Once your product reaches a certain level of maturity, gamification becomes the belle of the ball. Every executive will ask the design team to look for ways to "gamify"...all the things. But what does it mean to apply gamification and is it effective? Gamification means applying game-design elements and principles in nongame contexts to engage and motivate people to achieve their goals. Which sounds really good, especially in the context of belonging. People are internally motivated to play games, which should lead to better outcomes for your product. Right? Right?! Well, much like a Facebook relationship status, it's complicated.

Game-design techniques work well in games because you are willfully engaged in a recreational activity with the expectation of entertainment and connection as the end goal. So, if you decide to play a game of pickleball or go slay some monsters in a dungeon, you have a reasonable expectation that everyone is there voluntarily to have some fun and willing to play by the rules to enable that fun. However, it is unlikely that you would want to promote that same sense of belonging and engagement for, say, renewing your passport. There is no gain for your user or for your product in this type of scenario. Similarly, if your product runs reports on financials once a quarter, accuracy and efficiency will be valued far more than engagement or competitiveness. You'll often hear a request to make experiences "sticky," meaning to make them something people want to do over and over again and stay with. Just remember – no one wants their hammer to be sticky (that's just gross), and MOST tasks require hammers (a.k.a. tools that help you do something specific faster).

Scenarios in which gamification makes sense:

- Areas in which there is a benefit to the user in increasing regular engagement, such as professional communities, loyalty programs, getting an annual physical, or training/learning experiences

CHAPTER 4 BEING PART OF SOMETHING

- Workflows that logically build toward an achievement, such as profile completion or performance/skills training within a career progression

- Competitive scenarios with demonstrable benefits to the users in comparing themselves with others, such as sales quotas and support call satisfaction

- Games!

Scenarios in which gamification does NOT make sense:

- Tasks that are part of someone's everyday job and require a "hammer" such as running reports, analyzing data, building a website, and writing a paper

- Areas that are required due to regulation such as paying taxes or renewing a passport

- Workflows that are necessary to fix a problem such as opening a support ticket and repairing your roof

- Scenarios that happen once or very infrequently such as buying a house, installing an enterprise security solution, and having surgery

How to Game Belonging

If you've taken a look at your workflow or scenario and decided it would benefit the user if you gamified it, you might be wondering "now, what?" Well, there are a lot of potential ways to promote a sense of belonging and engagement with your designs (assuming that is relevant and important for your goals).

CHAPTER 4   BEING PART OF SOMETHING

Warning   It's important to choose the right technique for the right scenario or you risk alienating your audience – for example, using leaderboards when there's no competitive aspect to your engagement will not simply have no effect but could cause engagement to drop as users become annoyed with the extra visual noise and unwanted interaction.

The following sections outline some common game design techniques and the scenarios in which you might use them effectively.

Points and Scoring Systems

Points and scoring systems assign a value to accomplishments with your context. Giving someone points for their work can provide immediate feedback and a sense of accomplishment.

A good scoring system balances challenge and reward, providing you with a sense of accomplishment without making things too tough. Outside the gaming context, point systems can be challenging – providing meaningful rewards is essential, which can be difficult to do in a work setting. The feedback aspect of points systems is particularly key as it helps users understand how their actions impact their scores. Many food establishments and retailers effectively use this form of gamification to encourage repeat customers, letting you earn points toward a gift or coupon.

Badges/Medals

One of the most common gamification techniques is badges. This is not to be confused with badges on the icons on your phone – those badges are tapping into the cognitive biases we mentioned in Chapter 3. Gaming badges are visual representations of accomplishments that users can

CHAPTER 4   BEING PART OF SOMETHING

earn by completing specific actions or reaching milestones. You might get badges for finishing a level, defeating a boss, finishing your profile, providing your email, finishing a survey, or completing a language lesson ten times on successive days. Badges are a visual reminder that you have accomplished something and an encouragement to accomplish more.[16] Badges are effective when you have specific, meaningful things your user can accomplish or strive for. Membership organizations benefit from this type of gamification as the badge can be associated with the number of volunteer hours, having attained a certification, or similar aspect that is connected to status or accomplishment within the group.

---

[16] Balci, S., Secaur, J. M., & Morris, B. J. (2022). Comparing the effectiveness of badges and leaderboards on academic performance and motivation of students in fully versus partially gamified online physics classes. *Education and information technologies*, *27*(6), 8669–8704.

95

CHAPTER 4   BEING PART OF SOMETHING

Leaderboards

Leaderboards are popular in games because they let you foster a sense of competition which can in turn motivate your players to strive to achieve more and improve their performance. Sales groups often use this technique, posting rankings of sales people based on the volume or value of their completed sales. While this level of social comparison can be useful for games in and of itself, when applied outside the world of games, it can fall short. Motivation in business is different because you are already being compensated through your salary, commission, or bonuses. To be effective in a nongame setting, leaderboards must be coupled with a meaningful time-bound reward – for example, a trip to the Bahamas awarded to the salesperson who made the most sales in the past calendar year or a new iPad for the student that made the highest grades in a school year.

Designing for Trust

One of the issues with AI is trust. Due to an LLM's propensity for "hallucinations" (asserting items as fact that are untrue), it can be difficult to trust AI results. This fact is critical when choosing to integrate AI into your experiences to promote belonging because it may very well backfire. A great example of this is Father Justin, an AI construct intended to provide quick advice and insight to visitors to the site Catholic Answers. Father Justin erroneously offered some outrageous advice to petitioners including that they could baptize their babies in Gatorade and that it was OK to marry your brother.[17] This blatant violation of trust, while amusing, is precisely the opposite of what the designers intended.

---

[17] https://www.pcgamer.com/software/ai/catholic-media-ministry-defrocks-awol-ai-priest-[...]n-gatorade-and-that-sure-it-can-totally-perform-your-wedding/

CHAPTER 4   BEING PART OF SOMETHING

What is trust? At its core, trust is the level of confidence you have in another entity to fulfill your expectations. Trust is a core part of belonging – you must trust the organization, group, etc. you are associating with to uphold your belief in them. If your local tea group suddenly decided to go fossil hunting instead of trying the latest Darjeeling, you would likely find this a violation of your trust which would cause you to reevaluate your association with that group (unless you are Dr. Baker, in which case it would be expected). A breach of trust doesn't have to be extreme to have serious consequences in communities. For a retailer, it could be as simple as taking three days to deliver your goods instead of one. For a membership organization, it could be failing to have enough speakers for your conference. But trust violations can be even smaller – in things we

CHAPTER 4 BEING PART OF SOMETHING

might think of as insignificant as designers but represent an important experience for our users.[18]

Think about the last time you used an application in which the design didn't follow standard design principles – perhaps they put the button to execute the sale at the top of the page rather than the bottom, or they moved the navigation to the right side of the page rather than the left, or maybe they colored the "go" button red instead of green. Taken in isolation, these design decisions don't break things and might have been implemented by the designer to be edgy or exciting. But how do they make you feel? For the majority of the population, violating expectations for design elements intended to help you complete a task causes a loss of trust. And if you can't trust the interface you're using, why would you use it?

Nurturing Belonging Through Community

One way in which belonging is manifest is through shared membership in groups that focus on consumption of a brand or experience. For example, one of us may or may not have camped outside for a day and a half in order to purchase tickets to see *The Phantom Menace* on its opening day (and it should be noted that the same individual knows he will never live that down). Whether they be *Star Wars* nerds, Trekkies, sneakerheads, or die-hard consumers of Apple products, brand loyalty both creates and fulfills a sense of belonging for the individuals in those groups.[19]

Though something as iconic as Apple or the sci-fi adventures in a galaxy far, far, away may seem ready-made for the formation of groups, studies have shown that even less exciting or innovative brands can

---

[18] https://medium.com/@cassininazir/the-shape-of-trust-ac913a227a13
[19] https://www.bbc.com/worklife/article/20230815-why-some-brands-reach-cult-status

CHAPTER 4   BEING PART OF SOMETHING

institute a sense of belonging among customers through the establishment of online communities that provide opportunities for members to discuss products, share their experiences, and exchange ideas.[20]

Of course, establishing a community can be harder than it sounds. Designing a safe space for exploration of a shared interest tends to be a bit of a catch-22 scenario. You have to have a certain number of members for people to reap a benefit that makes the community worth joining. But you can't get that critical number of members without having members already in the community. Effective community building requires careful research into the target audience, what benefits you can offer for being part of the community (not what benefits your company will reap from the community – although that is important as well from a cost–benefit standpoint), and how you are going to maintain and grow that community through new content or benefit offers.

Cults vs. Communities: The Dark Side of Belonging

Belonging to a community is not all goodness and light. There can be a dark side to associating too strongly with any particular group. Take for example the connectedness experiments of Alex Pentland.[21] They studied a group of financial traders that used a platform that lets you copy other people's trades. They wanted to see how performance varied based on how much social interaction you had with other traders. What they found was interesting – those that chose to be isolated (did no copying) and those who copied exclusively from a set group (an echo chamber) did OK, but those who diversified their networks, copying from many different

---

[20] Royo-Vela, M., & Casamassima, P. (2011). The influence of belonging to virtual brand communities on consumers' affective commitment, satisfaction and word-of-mouth advertising: The ZARA case. *Online Information Review, 35*(4), 517–542.
[21] Social Physics, Alex Pentland.

CHAPTER 4   BEING PART OF SOMETHING

groups of individuals, outperformed the rest by almost 30%. That is, overidentification with a particular group was on par with isolation. We see this effect in other areas as well.

ECHO...ECHO...ECHO...

What is an echo chamber? An echo chamber is when you have a group of people who all reinforce the same opinions with each other – whether or not those opinions are correct. This reinforcement makes disagreeing or innovating difficult and plays on one of those biases we mentioned earlier – confirmation bias. We create our own echo chambers when we search the Internet looking only for facts that reinforce what we want to believe. Something to remember the next time to try to find out if drinking wine will help you lose weight.

Overidentification can also lead to a stifling of the individual. In the 1950s, Solomon Asch conducted a series of experiments at Swarthmore College that became the foundation for our understanding of how group dynamics can influence people's behaviors. In his early studies, Asch would tell research participants that they were being tested on their visual acuity. These participants would be led (last) into a group setting where there would be a variable number of other "participants." We used quotations there because unbeknownst to the *actual* participants, everyone else in the room was actually a plant by the research team (i.e., confederates) who would respond to answers using a predetermined script.

The study, which was actually designed to measure conformity, would present all of the participants with an initial line (on the left). They were then shown three test lines (right, A–C) and asked to judge which of the three lines was closest in size to the one they were initially shown. In the first few rounds, the confederates, who would always give their answers

100

CHAPTER 4 BEING PART OF SOMETHING

before the real participant, would give the correct answer (i.e., "C"). In later rounds, however, the confederates would unanimously start to give answers that were unabashedly incorrect (i.e., "A"). These critical trials were then tested to see how the actual research participant would respond. Would he (yes, it was always "he"...blame it on the 1950s) give the obviously correct answer and break with the egregiously wrong, but unanimous, answer of the group or give the answer that wouldn't make him look different?[22]

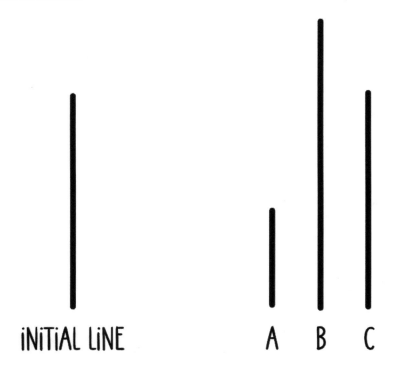

---

[22] Bond, R. (2005). Group size and conformity. *Group processes & intergroup relations*, 8(4), 331–354.

CHAPTER 4 BEING PART OF SOMETHING

Asch found that roughly one-third of participants would give an incorrect answer (i.e., conform to group pressures) at least once during the trials. He also found that if there were only one other person, conformity wasn't as likely as it was when there were three other individuals giving a wrong answer.

So what do these data have to do with design in the 21st century? The applicability is less about what is crafted than the processes via which original ideas are considered, vetted, and decided upon. Collaboration, though paramount to the design process, can result in groupthink if individuals worry about whether they will stand out, feel unsure about voicing dissenting opinions, or aren't secure with their place in an organization. In fact, groups that include voices that go against the grain have been shown to make better decisions and enjoy more creativity and innovation.[23] Thus, though group cohesion is important, it is critical that leaders foster an environment where all voices are able to be heard.

Recap

- As humans, we are wired to identify with others and create a group identity that leads to a sense of security and connection.

- Designers can take advantage of this wiring to create experiences that recreate a sense of belonging to improve the overall impression and desirability of the product.

- Gamification is one way to encourage a sense of belonging, but it has to be applied judiciously to be effective.

---

[23] Jetten, J., & Hornsey, M. J. (2014). Deviance and dissent in groups. *Annual review of psychology*, *65*(1), 461–485.

CHAPTER 4   BEING PART OF SOMETHING

- Trust is an aspect of belonging that we should consider when designing – it is difficult to gain and easy to lose.

- Not all belonging is good! Overidentifying with a group can lead to echo chambers and reduce diversity of thought.

Before You Go...

In 1962, *Candid Camera* aired an episode called "Face the Rear" in which unsuspecting people got on an elevator then everyone else who got on afterward (all members of *Candid Camera*) faced backward.[24] A surprising number of people, when confronted with an elevator full of people facing the back, chose to also face the back rather than be "different." At one point, the plants in the elevator (like the confederates in the Asch study) begin a slow rotation, first facing the right side, then the back then the left side, and, sure enough, the unwary person turned with them. This instinct is the same one that leads you to sit in the same seat every day in a lecture, even though they are not assigned, or to take the same parking spot at work. It's important to be aware of this tendency in ourselves, not because it is, in and of itself, bad but because it can lead us to open ourselves up to manipulation. Dr. Robert Cialdini warns about the perniciousness of overidentification, the tendency to think in terms of "we" and "them":

> *When asked to rank-order a waiting list of patients suffering from kidney disease as to their deservingness for the next available treatment, people chose those whose political party matched theirs.*[25]

---

[24] https://www.youtube.com/watch?v=7YOKTOajZBw

[25] Cialdini, Robert B. Influence, New and Expanded: The Psychology of Persuasion (p. 371). HarperCollins. Kindle Edition.

CHAPTER 5

Defining the Box

JLR: So, one of my students asked me if you could tell if someone was insane by what they see in clouds.

RB: What did you say?

JLR: I asked what they saw and then confirmed for them that only a crazy person would see what they had seen.

RB: Hmmmm.

JLR: OK, so I didn't do that (but I wanted to do so), but what I did do is use it as a starting point for a class discussion of how the reality that each of us experiences is truly unique yet is subject to the same limitations. I think the class was both comforted by this and...unsettled.

RB: That sounds about right. What *did* the student see in the clouds?

JLR: Sonic the Hedgehog brewing beer.

RB: Perhaps you should rethink the answer you gave them.

Understanding the Limitations of Human Experience

Take a moment and think about all of the information about your environment that you are currently taking in through your senses. There are the words of this book, in addition to whatever other sensory information you can see when not looking at the page. There are the sounds in which you are immersed. Perhaps you are listening to music. Perhaps you are even listening to *good* music (i.e., The Grateful Dead).

© Jerome L. Rekart and Rebecca Baker 2025
J. L. Rekart and R. Baker, *Designing for Human Intelligence in an Artificial Intelligence World*,
https://doi.org/10.1007/979-8-8688-1418-1_5

CHAPTER 5   DEFINING THE BOX

Given our modern reliance on machines, there are probably any number of other sounds that you are experiencing at the same time. Car horns, alarms, the gentle hum of a refrigerator, the ping of cell phone alerts – although you may not be paying attention to these (more on that later in Chapter 6), if you listen you can hear them. If you move your arm, you may feel the soft rustle of a sleeve on the fine hairs of your arm. In addition to those three *physical* senses, which detect either waves of light (vision) or pressure waves (hearing and touch), there are others as well. The chemical senses of smell and taste. Our bodily knowledge of where we are in space (e.g., Are we sitting up? Accelerating or decelerating?) and our sense of time.

Despite the richness of sensory information that we encounter in even the most mundane of contexts, there are limits. Our sensory processes, which feature an incredibly acute sense of vision and decent hearing, are not even at the limits of what is experienced on this planet. Our visual system, for example, is limited to just a small band of light that occurs between 380 and roughly 700 nanometers (nm; i.e., lightwaves that have lengths measured in the billionths of meters or roughly 1.4 hundred thousandths of an inch to 2.8 hundred thousandths). Though this is enough to give us the richness that constitutes everything from the vibrant red of a strawberry to the deep violet of a night sky, there are animals who can see further into the red end of the spectrum (i.e., infrared), like pit vipers, and those that can see more than 700 nm, which is in the ultraviolet region (e.g., honeybees). Thus, if we think of our experience of the visual world, it would be akin to knowing that there are radio stations on the FM dial that can tune into music at less than 95 MHz and greater than 100 MHz but never being able to listen to them (without the help of machines). And this is for the premier sensory function that we possess (vision). For the others, which are decidedly less acute and rank much lower in terms of the animal kingdom, our richness of experience is comparatively destitute.

CHAPTER 5   DEFINING THE BOX

Sensation + Perception = Experience

Though I don't think either of us would claim membership in groups of individuals who describe themselves as "pepper heads," both of us do enjoy the taste and experience of hot peppers. The fact that there is great variability in how people respond to capsaicin-containing foods like peppers provides an ideal platform to discuss how the same sensory inputs can be perceived differently across individuals.

As most of us know, the sensation of taste occurs when the various molecules that enter the mouth activate the various taste receptors found in the buds on our tongue. Unlike the tastes of salty, sweet, savory (i.e., umami), sour, and bitter, the spiciness of hot peppers occurs through the activation of a different type of receptors, which are also located throughout the mouth (and elsewhere on the body). These receptors respond to various irritants and produce a burning-type sensation in response to chemicals like capsaicin, which produces the "heat" in hot peppers like jalapenos, habaneros, and ghost peppers, and piperine, which is what makes the condiment pepper (of salt and pepper fame) taste, well…peppery. The activation of these receptors occurs in a somewhat linear manner, with hotter peppers releasing more capsaicin, which in turn activates more of these receptors. This means that the *sensation* of the pepper's heat is directly related to the quantity of

CHAPTER 5   DEFINING THE BOX

capsaicin regardless of who is eating the pepper. Thus, a ghost pepper, which has roughly 1,000,000 Scoville Heat Units (SHUs; which is how a pepper's heat is measured), will always be sensed as being hotter than a jalapeno, with its roughly 8,000 Scoville Heat Units.[1]

However, how the heat of different peppers is *perceived* will vary from individual to individual.[2] For some, a jalapeno will produce pain and unpleasantness, whereas for others (like both of us), they are actually perceived as quite mild. Indeed, even hotter peppers like serranos (10,000–23,000 Scoville Heat Units) and some (though not all) sauces containing habaneros (as much as 500,000 SHUs) are for some (like us) enjoyable. This is not to say that they aren't hot but that the heat is perceived as pleasurable and tasty. Thus, the perception *of the* sensation is what produces our individual experiences.

The same relationship between sensation and perception holds true for our other senses as well. For vision and hearing, our brains apply complex heuristics and look to our memories for hints about deciphering what we are seeing or hearing at any given point in time. This is why when you look at the shapes of clouds, you may perceive something different than someone else who may be looking at the exact same thing. This also explains why we sometimes "see" things that aren't there, as is often depicted in that old trope about children seeing monsters in piles of laundry scattered throughout their room (which is why they should always be folded and put away promptly!).

---

[1] https://www.nist.gov/how-do-you-measure-it/how-do-you-measure-heat-pepper

[2] Siebert, E., Lee, S. Y., & Prescott, M. P. (2022). Chili pepper preference development and its impact on dietary intake: A narrative review. *Frontiers in Nutrition*, 9, 1039207.

CHAPTER 5 DEFINING THE BOX

Similarly, the viral clip that divided people into the "yanny" hearers vs. the "laurel" hearers was due to an ambiguous audio file being perceived differently by people – despite the sound file being the exact same for all.[3]

IN THE EYE OF THE BEHOLDER

Art is another example of the importance of perception and context. How we interpret what we see may be very different – the same visual input of the Mona Lisa may appear with a smile or a smirk or the auditory input of Metallica's "Nothing Else Matters" can be heard as depressing or rebellious. Your memories, experiences, and context deeply affect your perception of the world around you and how you interpret it. And not just your personal context but your belief or knowledge of the artist's context. In a recent study on factors affecting the aesthetic perception of art,[4] it was found that providing negative social details about the artist caused people to think the art was of lesser quality, regardless of the actual quality of the art.

AI, on the other hand, usually doesn't "see" or "hear" anything, but rather most large language models will look for patterns across what humans have *described* seeing or hearing and use that in place of actually sensing the information on its own. Thus, the perceptual output (i.e., what it generates) will be skewed by the majority of people, and it is highly unlikely that it would ever introduce any novelty. Put differently, if an image of a cloud were examined by people, and their results were indicated online, it would be limited to only what those people saw.

---

[3] https://www.npr.org/sections/thetwo-way/2018/05/16/611701171/yanny-or-laurel-why-people-hear-different-things-in-that-viral-clip

[4] Kaube, H., Abdel Rahman, R. Art perception is affected by negative knowledge about famous and unknown artists. Sci Rep 14, 8143 (2024). https://doi.org/10.1038/s41598-024-58697-1.

CHAPTER 5   DEFINING THE BOX

Sensation and a Preference for Detecting Change

As you sit (or stand or somersault) while reading this passage, you are likely not aware of the pressure of the various articles of clothing that you are wearing (assuming you have anything on...you know what, let's assume that you do). Though, if we move around, we may become cognizant of sleeve folds or roughly sewn seams or other textural variations of our fabric tops and bottoms, we spend most of our days numb (though thankful) to the fact that we *are* in fact wearing clothes. This example illustrates how our sense of touch, just like our other senses, is more attuned to changes than constancy. This phenomenon is called sensory adaptation by scientists who study how our senses work: psychophysicists (note that if there were a space between the "o" and second "p" – "psycho physicists" – then it would describe someone who tries to figure out how to change the moon's orbit in order to enslave mankind).

We likely evolved a system that responds to changes in stimuli, like most adaptations, because it facilitated the survival of our ancestors hundreds of thousands of years ago. What this means for us in modern times is that we have systems that do a fantastic job – within limits – of responding to the outside world but are not designed or equipped to monitor or respond to the *status quo.*

The ways in which we do detect changes are dependent upon the size of the change relative to the resting state. Put differently, if we are holding two one-pound bags of rice and add one grain to either, we will not be able to discern which bag contains the additional grain because the change makes up such a small percentage of the overall quantity. Similarly, if a noise is incredibly loud (130 db) and then the volume is increased slightly (say by 10 db), then we are less likely to notice this change (~7.6%) then if someone who was speaking at 50 db (normal conversational level)

CHAPTER 5   DEFINING THE BOX

suddenly increases their volume to 60 db (a change of 20%). In fact, there are well-established fractions via which auditory and visual changes can be detected, which are roughly constant for humans (e.g., Weber fractions).

From a design perspective, it is important to note that shifts in the parameters of visual stimuli (called perceived animacy) need to be robust enough that the typical viewer will recognize the change. These parameters apply to animations as well, given that we often use color shifts to simulate motion in otherwise static objects. The speed, magnitude of change, and whether the motion appears to be self-driven (vs. driven by interaction with another object) can all affect how noticeable an animation is and hence how effective it can be in capturing attention.[5]

Time Travel: Perception and Time

Although time is an integral part of our everyday experience, there is still much that is not understood about how we internalize its passage. Humans and other animals do have specific genes and neurons that respond to sunlight and allow us to "set" our internal clocks in a manner that aligns with the rising and setting of the sun.[6] The rhythms engendered by these processes are referred to as circadian, which nicely breaks down from the Latin into "about" (*circa*) "once a day" (*dian* derived from *diem*). Though circadian rhythms are important for everything from sleep regulation to mental health, they are limited in that they don't really allow us to tell time per se. Rather, they provide us with more of a sundial that has the numerals rubbed off (e.g., "It is mid-afternoon now") than a clock tower (e.g., "It is 3:57 PM") in terms of how they enable our intuition of the passage of time.

---

[5] https://www.nngroup.com/articles/animation-usability/

[6] Patke, A., Young, M. W., & Axelrod, S. (2020). Molecular mechanisms and physiological importance of circadian rhythms. *Nature reviews Molecular cell biology*, *21*(2), 67–84.

CHAPTER 5   DEFINING THE BOX

Thus, our internal clocks give us the gist of what time it is, but like a sundial, they do not provide us with a reliable way of determining how much time has passed. This is an important distinction as our perception of the passage of time is a fundamental facet of our lived experiences. We judge time to determine if we've waited long enough for a web page to load and whether we have enough time to throw in a load of laundry before we have to leave for a dinner date. Even in our everyday vernacular, we use terms that speak to the subjectivity of our perception of time.

Verbs like "dragged" and "flew" denote time that seems to be passing slowly or quickly, respectively. And yet, despite the functional importance of knowing how much time has elapsed as humans go about their days, neuroscientists have yet to agree upon the specific pathways[7] that must be responsible for the internal stopwatch of our minds. Strikingly, although

---

[7] Naghibi, N., Jahangiri, N., Khosrowabadi, R., Eickhoff, C. R., Eickhoff, S. B., Coull, J. T., & Tahmasian, M. (2024). Embodying time in the brain: A multi-dimensional neuroimaging meta-analysis of 95 duration processing studies. *Neuropsychology Review, 34*(1), 277–298.

CHAPTER 5   DEFINING THE BOX

time is experienced by all of us, we do not have sensory organs that allow us to interact with it in the same way that our other senses enable our movement and passage through the physical dimensions of our world.

As with the perception associated with our standard sensory repertoire (e.g., vision, smell), our perception of time is both subjective and oftentimes inaccurately represents what has actually occurred. In fact, some of the most reliable distorters of our perception of the passage of time are our emotions.[8] This should not be surprising as we have all experienced the...seemingly

> ...glacial

> ...passage

> ...of

> ...time

> ...when

> ...we

> ...are

> ...bored.

Consistent with this, so-called negatively valenced emotions, like boredom, result in an overestimation or expansion of our perception of how long things take. Indeed, the more negative the situation, the more time seems to slooooooooooooooooooooooooooooooooow down and stand still. Interestingly, positive scenarios are also perceived as taking longer than neutral ones but pale in comparison to those that are aversive or unwanted. On the other hand, situations that are energizing (which in psychological terms is referred to as "arousing") result in a perception

---

[8] Droit-Volet, S., & Meck, W. H. (2007). How emotions colour our perception of time. *Trends in cognitive sciences, 11*(12), 504–513.

CHAPTER 5   DEFINING THE BOX

that time moves more rapidly, a phenomenon that has the sci-fi-sounding name of "dilation" of time.[9] Although the interaction effects of valence (i.e., is it positive or negative) and activation can be complicated, a useful rule of thumb is to design toward *activating* positive vibes rather than merely minimizing negative ones.

NOT RELATIVE

If I might be permitted a bit of a soapbox here, our variable perception of time is often attributed, erroneously, to Einstein's theory of relativity. Comparing putting your hand on a hot stove vs. putting it on a hot person (*Deep Blue Sea*, paraphrased, LL Cool J) is not an appropriate or even apt application of this most famous of theories, aside from the fact that the speed of light can be measured identically in both situations. So please – tell your friends. Friends don't let friends do bad physics.

These results have direct ramifications for the overall layout of site design. In the "obvious column" is to remove as many aggravations as possible and optimize page loading times because the negativity associated with loading delays and lags will *seem* to take even longer than they actually are...with an undesirable effect that the longer someone is waiting, the more frustrated they will become (i.e., increased negativity), which in turn will result in greater likelihood that their perceptions of the delay will be *even longer* than it may be in reality.

---

[9] Cui, X., Tian, Y., Zhang, L., Chen, Y., Bai, Y., Li, D., ... & Yin, H. (2023). The role of valence, arousal, stimulus type, and temporal paradigm in the effect of emotion on time perception: A meta-analysis. *Psychonomic Bulletin & Review, 30*(1), 1–21.

CHAPTER 5 DEFINING THE BOX

In the "not-so-obvious column" of how to use findings about the interplay of emotion and the perception of time are some simple design tweaks. Of course web pages cannot load instantaneously, and even if they were designed to do so under optimal conditions, users will be accessing a given site through devices and using bandwidths that are less than ideal. Thus, minimizing frustration during web load times is a critical piece of good site design. In fact, studies have shown that not only can the introduction of colors and animations to loading screens minimize frustration, when they are seen as fun or whimsical, they actually decrease the perceived amount of time spent waiting.[10]

Another facet of temporal perception that bears examination here is the finding that one's subjective appraisal of how long something lasts (i.e., a stimulus, such as a light, tone, or even electrical shock) is directly related to the relative strength of the stimulus.[11] Thus, bright lights are seen as lasting longer than dim lights, rapidly moving objects are seen as persisting for longer periods of time than slower moving objects, etc. The fact that this "more (property of a stimulus)=longer (presence of the stimulus)" has been replicated across numerous studies, and with everything from brightness to the number of objects being shown, their relative size, volume, and weight (among others) should be considered when thinking about the design and elements that may accompany waiting or periods where users or consumers aren't meant to be engaged or you would prefer that they perceive time as moving faster (i.e., shorter durations) than it actually does.

---

[10] Cheng, A., Ma, D., Qian, H., & Pan, Y. (2024). The effects of mobile applications' passive and interactive loading screen types on waiting experience. *Behaviour & Information Technology, 43*(8), 1652–1663.

[11] Matthews, W. J., & Meck, W. H. (2016). Temporal cognition: Connecting subjective time to perception, attention, and memory. *Psychological bulletin, 142*(8), 865–907.

CHAPTER 5   DEFINING THE BOX

AI Making Things Easier...or Not?

AI has been touted as a great equalizer. Proponents suggest using this technology to provide adaptive interfaces for those that have special needs such as vision or hearing impairments.[12] Certainly, AI has great promise in helping people do things they might not otherwise be able to do, from creating art work and prose to programming applications and writing letters.

However, as with all things, we need to look deeper. Is the interface for AI itself accessible and usable? Many platforms now charge a fee to be able to access the technology, creating an instant financial barrier for many. The algorithms are far from foolproof (see previous on hallucinations) and require the user to have the ability – the context and knowledge – to be able to check their results and recognize when they are good or bad. And of course, the interface for many AI engines is at best spare and at worst confusing to the average reader. AI can easily become elitist, losing the promise of accessibility altogether.

While we've talked a lot about the limits of human perception and how this affects our ability to design for humans, we think it's important to note that AI also has perceptual limitations. We often treat AI as a rather pedantic coworker who tirelessly answers questions for us at a whim and whose only fault lies in an inability to fetch us a good cup of coffee. But what are the limits of AI? It has no eyes or ears. It can't touch or smell anything. As we talked about earlier, perception is equal parts context and input. So, for AI, like OI, its context will hamper or enhance its ability to perceive input. For example, consider Google's Gemini creating images of Nazi soldiers of color.[13]

---

[12] https://jakobnielsenphd.substack.com/p/accessibility-generative-ui
[13] https://www.nytimes.com/2024/02/22/technology/google-gemini-german-uniforms.html

CHAPTER 5   DEFINING THE BOX

Or this charming AI-generated alert Dr. Baker received on LinkedIn:

 Overstimulated at work? Identify your triggers to strategize solutions. Here's how.

Overstimulated? What am I, a toddler?

Both of these are examples, not of hallucinations, but of lack of context on the part of the AI which leads to a poor perception of the request and an amusingly off-target result. A lot of work is being done in this area to help systems have better context, including providing curated training sets that provide greater focus for the systems learning from them. As systems become more advanced, they start to include cultural queues that inform so much of our perceptual interpretations.

It's important to take these limitations into account when working with AI. Contextual cues we take for granted may be missing from the training sets used by these systems which can lead to…variable…results. To do this effectively, ensure your prompts are specific in context and output, includes the purpose which you are looking to fulfill, and uses examples to help guide the result to achieve the right tone and format. Here's some good (and bad) examples for prompts that help provide context:

CHAPTER 5   DEFINING THE BOX

Goal	Bad Prompt	Good Prompt
Get a raise from your boss by making a case in email.	Write an email that will get me a raise.	Write an email making the case that I should get a 5% raise, taking into account that I am a sales associate working in a Toyota car dealership in Chicago, IL, and that I have exceeded my sales quotas three times in the last six months. Use a professional tone and keep the length of the email no longer than four lines.
Create a D&D adventure	Make a D&D campaign that will be fun.	In the style of Terry Pratchett's *DiscWorld*, write a third level D&D adventure to rescue someone from being sacrificed to a volcano that can be completed in two 3-hour sessions with at least three encounters with monsters, two puzzles to solve, and five nonplayer character encounters.
Write a blog post about the relationship between The Grateful Dead and Metallica.	Write a blog post that compares The Grateful Dead and Metallica.	Write a blog post using a tongue-in-cheek tone like *Etiquette and Espionage* by Gail Carriger, comparing the musical works from the bands The Grateful Dead and Metallica. Include pertinent examples and performances to compare.

Working In and Out of Limits

Why take the time to understand our limits as humans (other than to have some interesting debates over cocktails)? Designing for our very human limitations means designing for success. By ensuring the object, application, what have you will operate within the capabilities of the human who must use it, you ensure that it will enable success rather than

CHAPTER 5 DEFINING THE BOX

thwarting it. And, honestly, who wants to make something that's too hard to use (except maybe tax forms – but that's another story)? By creating things that are easy to use and enable people, we make using those things joyful. And that, in turn, makes the people who use them happy – both with themselves and with the tool you've provided them with.

WAIT, YOU'RE WHERE?

While the limitations that come from being human are important, it is also key to remember the limitations that come from your user's environment. Consider how you design, say, a mobile application. You are probably sitting upright, working on a laptop or maybe a large extended monitor in a well-lit office or coffee shop, pulling in pieces into your design frame that is sized just so for your mobile device. When you test it, you click through with your mouse and nod to yourself in satisfaction in how easy to use this workflow is. Except... no one uses an app on their phone like that. People use mobile apps hunched over in line with the phone balanced in one hand and a toddler in the other using a thumb to try to select the item, lying in bed in the middle of a dark room squinting at the read out, or sitting in a small chamber attending to, um, more biological imperatives. Good design works everywhere — even those places you'd rather not think about your app being used.

Forgetting Limits: Rage Clicking

A classic example of what happens when you do not design with your human limitations in mind is rage clicking. Rage clicking occurs when users rapidly and repeatedly click on an element of a website or application out of frustration. This behavior often signals that something is going wrong in the user experience, leading to confusion, impatience, or anger. Rage clicking can nearly always be tied back to failing to design

CHAPTER 5 DEFINING THE BOX

for the "box" (a.k.a. limitations) of the program and the people using it. A great example of rage clicking is in the movies when the protagonist is being chased and needs to enter an elevator to escape. They push the button repeatedly and with greater violence while the tension builds until the doors finally open and they rush into the elevator, only to repeat the same clicking as they try to get the elevator to close. Examples of things that lead to rage clicking outside of the movies include

- **Slow Load Times**: If a web page or specific elements like buttons, videos, or links take too long to load or respond and do not include a progress indicator, users may start clicking repeatedly, thinking their initial click didn't register.

- **Broken Links or Buttons**: When a clickable element appears to be interactive but doesn't work as expected, users may try clicking multiple times.

- **Lack of Immediate Feedback**: If there's no visual or auditory feedback when a user clicks on something (like a button change, animation, or sound), they may assume their click wasn't recognized, prompting them to click again.

- **Small or Overlapping Touch Targets**: If buttons or links are too small or too close together, users may accidentally click the wrong one, leading to repeated attempts to click the correct target.

- **Stress or Time Pressure**: Users who are in a hurry or stressed are more likely to become frustrated by any delay or complication.

120

CHAPTER 5 DEFINING THE BOX

Rage clicking is a clear sign that the user experience is not meeting expectations. It often results from unresponsive elements, poor feedback mechanisms, misleading design, unexpected behavior, or slow performance – all items that can be identified as part of the environment and the perceptual limitations of human beings. By identifying and addressing these issues, designers and developers can reduce frustration, improve usability, and create a smoother, more satisfying user experience.

Designing for Everyone: Universal Usability

Many designers, developers, and product people think of accessibility guidelines as being limiting and difficult and, most importantly, only necessary for people with disabilities. This attitude, unfortunately, ends up being self-fulfilling and creates products that are inaccessible with awkward workaround for people with specific challenges. It fails to recognize that everyone is disabled at times. I used to have perfect vision, but as I've gotten older, I need reading glasses to be able to see most print. And before my morning cup of coffee, I can have trouble focusing and sometimes have to ask people to repeat themselves. But let's think of a more extreme, but realistic, example. Imagine for a moment you are designing the shutdown system for a nuclear power plant in case of fire. How does your user experience that system? The room is filled with smoke, limiting visibility, blinding you. Fire alarms are blaring everywhere, limiting your ability to hear anything – you are deaf to everything except the sound of your labored breathing. Your stress levels are through the roof as you search your brain frantically for the shutdown sequence and you're having trouble thinking clearly. If this system was not designed with accessibility in mind, it would fail a "normal" person.

Universal usability suggests a mindset which focuses on the problem to be solved for a diverse set of users, rather than the problem to be solved for a narrow set of users. That is, rather than trying to design the "normal" and treating those with disabilities as exceptions, it tells you to design for the broadest set of needs and treat "normal" as the exception. Let's look at some examples of successful universal usability design.

121

CHAPTER 5   DEFINING THE BOX

Curb Cuts

Curb cuts[14] – that small ramp that creates a break in the curb connecting the street to the higher sidewalk – are a ubiquitous sight these days. However, they didn't start appearing on our streets until 1945, spurred by the many disabled vets returning from the war. Originally intended to improve the lives of those with motor disabilities, curb cuts are now used by everyone – from the parent with a child in a stroller to the skateboarding teenager, to the busy business person on the way to the airport with their rolling luggage. As such, they are an excellent example of how a thoughtful design adjustment can benefit not just the target audience but everyone else as well.

Typewriters

Typewriters, while less common today than 20 years ago, have shaped the way we write and communicate with one another. While you may not find a typewriter in regular use today, the keyboard on your laptop was created based on this model. The first typewriter was not, however, invented as a faster way to create prose but rather as a help for a friend who was blind. Pellegrino Turri[15]created the typewriter for his friend, or perhaps lover, the Countess Carolina Fantoni da Fivizzano, who had gone blind, to enable her to be able to write letters in 1808. The typewriter used carbon paper (which Turri also invented) to ink the paper with the typewriter.

---

[14] https://www.carleton.edu/accessibility-resources/newsletter/curb-cuts-a-brief-history/

[15] https://thehistoryofcomputing.net/from-moveable-type-to-the-keyboard

CHAPTER 5   DEFINING THE BOX

LOVE IS BLIND

Some of the original letters from the Contessa still survive and were the inspiration for a historic fiction novel *The Blind Contessa's New Machine* by Carey Wallace. In the novel, ignored by her family and husband, the Contessa turns to Turri as the only one who believes she is going blind. He devises the typewriter so that she may continue to write letters after her vision has faded which leads the two to a steamy affair (in the grand scheme of romantic adventures "hey babe, I invented a typewriter for you" is definitely top ten). It is worth pointing out that there is no evidence (from the Countess' surviving letters or otherwise) that the two actually had an affair, but what a great story!

Recap

- Although we may all live in the same world, our perceived experience of that world will be different.

- Our perception of time is influenced most strongly by our emotions, with negative situations seeming to slow things down and activating or exciting situations speeding them up.

- Time perception is also influenced by the perceived magnitude of a stimulus, meaning that louder, brighter, faster, etc. objects are experienced as lasting longer than softer, duller, and slower ones.

- AI is also affected by limitations – designing prompts must take into account the limited context that AI has access to.

CHAPTER 5   DEFINING THE BOX

- Designing for human limitations is important to increase user satisfaction. Environmental considerations need to be factored into those limitations as well as cognitive limitations.

- Rage clicking is a phenomenon that happens when a user becomes frustrated with the responsiveness or design of the interface.

- Universal usability is a design approach that goes beyond accessibility by being inclusive rather than exclusive.

Before You Go...

As we've discussed, AI lacks context. A New Zealand supermarket experimented with an AI app that would allow shoppers to input ingredients they may see in the store or have in their cart, which the app would then turn into recipes. Unfortunately, the developers didn't account for the fact that folks might ask for, shall we say, "unique" combinations of ingredients. When prompted with the ingredients "ant poison" and "bread," the app recommended "poison bread sandwiches"; when "chlorine" and "bleach" were used as inputs, the app came back with instructions on how to make "aromatic water mix" (note: never, NEVER mix those two – it creates chloramine gas which IS deadly); and, perhaps, most disturbing, "carrots, celery, and human flesh" yielded "mystery meat stew," which specified that for the dish, one should use half a kilogram of said protein source![16]

---

[16] https://nextnature.org/en/magazine/story/2023/ai-supermarket-bot-suggests-serving-human-flesh-and-chlorine-gas

CHAPTER 6

Attention (or Lack Thereof)

RB: Have you ever thought about the fact that in English, attention is something that is "paid" rather than provided?

JLR: I know, weird, right? At least at first glance it is. I mean, I'm sure we could spin some kind of capitalistic, pro-free market reason for why that is, but other languages also have this notion of attention being a meaningful gift. In Spanish, for example, one "lends" attention ("prestar atención"), and in Hebrew, you put your heart to something in order to pay heed ("tsumet lev").

RB: Your little nerd moment actually has me wondering about other languages...

JLR: Yes, me as well, which brings home my point. As our everyday experiences teach us, attention is something that does require effort. This effort reflects the cognitive resources that we actually deploy to focus or concentrate on just one stream of information at a time – like how we can jump back forth from talking about attention to discussing how different languages deal with attention.

RB: You got me there. To double down on that point, we have all experienced that the more streams of information there are, the harder it is to focus on just one – there's just too much competing for your attention. This fact is HUGE when you think about design. Consider a banking app.

© Jerome L. Rekart and Rebecca Baker 2025
J. L. Rekart and R. Baker, *Designing for Human Intelligence in an Artificial Intelligence World*,
https://doi.org/10.1007/979-8-8688-1418-1_6

CHAPTER 6   ATTENTION (OR LACK THEREOF)

Chances are good that the designers tested it in a fairly quiet location with no distractions, probably sitting upright. But how do most people use a banking app? In line at the grocery store, in a car full of boisterous teens, at dinner with a large work group in a darkened restaurant or a loud sports bar...not exactly the "lab environment" the app was built for and tested in. So, how does our brain deal with all those distractions, and how can we concentrate on just the thing in front of us?

JLR: Maybe we can "spend" a little time talking about that?

RB: ...

The Bottleneck and the Pie

Although the title to this section seems like it could be a story from *Alice in Wonderland*, it actually refers to the two prevailing models of attention. Before we dive into the two models, and why they are both useful, we need to first get a feeling for the magnitude of information that we all process every moment. In 2006, Balasubramanian and colleagues produced an estimate (that holds to this day) of the amount of information that our retinae deliver to the brain every second. What they found is that roughly ten million distinct bits of information are transferred from our eyes to our brains every instant that they are open.[1] Coupled with our other sensory systems, we are looking at roughly 11,000,000 bits per second, or roughly the transmission capability of an ethernet cable of information streaming into our brains every instant (and just like what comes over the ethernet from the world, the vast majority of it is useless). And that is just what is

---

[1] Koch, K., McLean, J., Segev, R., Freed, M. A., Berry, M. J., Balasubramanian, V., & Sterling, P. (2006). How much the eye tells the brain. *Current biology*, *16*(14), 1428–1434.

CHAPTER 6   ATTENTION (OR LACK THEREOF)

coming in from the outside world. On top of all of that, there is our internal state, which ups the stream of information to be processed by our minds even more.

At any given moment, our internal state consists of all of the things that are on our mind – whether actively or that we've been contemplating. It is also our interpretation of the sensory information streaming in through multiple channels (i.e., eyes, ears, etc.) and the associations of those stimuli with memories we have of other times we've encountered them. Add to all of that our emotions (e.g., are we anxious?) and the status of our bodily state (e.g., are we hungry? Are we anxiously hungry: anxgry?), and you have a deluge of information. And yet, we rarely get overwhelmed.

In order to handle this onslaught of information, our brain has evolved multiple ways to parse through the attentional wheat and chaff. The first hypothesized way that it does this is to narrow the number of stimuli that are attended to in any given moment. In this model, all of the generalized features of stimuli are attended to (rather than all of the specific, minute details) and stored in working memory, with only an incredibly small subset of features (i.e., those minute details) rising to a conscious level where they become the focus of attention.

You can think of this as what happens with billboards we may see while driving at 65 mph on the highway. All we really process from them is the gist – the overall messages (drink more kiwi-flavored rum – yuck), some idea of the images (a kiwi, a bottle), and maybe some recognition of what is being advertised (New Zealand rum?). What we usually can't remember even moments later is what color eyes the model on the billboard had (even though said peepers may be over 15 feet tall).

The manner in which our brains determine which informational stream to focus on depends on several factors. For example, when the magnitude of a given stimulus rises meaningfully above the rest – think blaring alarms, flashing lights, and strong odors compared to the background sounds, sights, and smells – then they are preferentially allowed passage through what is called

127

CHAPTER 6   ATTENTION (OR LACK THEREOF)

an informational bottleneck.[2] In the absence of striking physical differences in magnitude (i.e., the difference in volume between a smoke detector going off and a whisper), it is the semantic features, which relate to how meaningful the information is to our current goals and situation, that determine what makes it through the bottleneck. And whatever doesn't pass through is stuck in the base of this informational bottle, where it will quite quickly fade into nothingness (at least as far as our minds are concerned – almost as if the bottle is filled with some highly corrosive substance, like hydrofluoric acid).

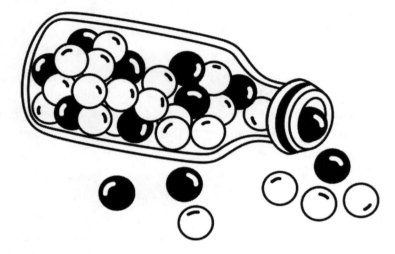

A great example of this process can be illustrated by conversations. The next time you have a conversation with someone, really focus on what they are saying. The way to know you are truly listening is by checking if you can answer a series of questions – while they are talking – such as

"What is this other person's intent?"

"What's their emotional state?"

---

[2] Lachter, J., Forster, K. I., & Ruthruff, E. (2004). Forty-five years after Broadbent (1958): still no identification without attention. *Psychological review, 111*(4), 880–913.

CHAPTER 6   ATTENTION (OR LACK THEREOF)

"What other information would be useful to this conversation?"

If you can answer these important social and contextual questions, then you'll know you've been actively engaged in the conversation. Congratulations, you've gotten the "gist" of what was said, why, and how (and were an active listener/good conversationalist, to boot).

Now, here comes the rub. The other task for you occurs at the end of this discussion in which you have been so actively engrossed. Think about the conversation you *just* had, and now pinpoint exactly what word they (the other person) used 15 *words* ago. Put differently, if that last sentence were part of an oral dialogue, then the 15th word back from "ago" would be the italicized "just."

Chances are excellent that you have no idea what the actual word used was (i.e., the physical stimulus from the real world) by the other person, but what you do remember is the broader concept that it was used to convey (i.e., the semantic features) of the conversation. Thus, in this model of attention, it is the features of the incoming information that determine whether or not it is attended to.

In the other prominent model of attention, the focus is less on how much information can be filtered through a bottleneck and more about the total number of resources required to process the information. In this model, which is often referred to as "divided attention" or "capacity" models, researchers try to determine how attention is carved up among tasks and stimuli.

With capacity models, our attentional resources have a limit (think of this at a mechanistic level – one person only has so many neurons that can be used to process information at a time[3]). If we envision that limit as an attentional "pie," then all of the stimuli that we're conscious of are

---

[3] Buschman, T. J., Siegel, M., Roy, J. E., & Miller, E. K. (2011). Neural substrates of cognitive capacity limitations. *Proceedings of the National Academy of Sciences*, *108*(27), 11252–11255.

129

CHAPTER 6   ATTENTION (OR LACK THEREOF)

slices of that pie, as are our internal state, emotions, and everything else that requires cognitive resources (remember that deluge of information). This means that if we're going to attend to something new, we must stop attending to something old OR must attend to the old information with fewer resources (smaller slices of pie).

Although we've just presented two models, they should not be seen as competing nor should one think that either you must champion one or the other (so, we'd put a hold on any "Team Capacity" or "Team Bottleneck" shirts you may be printing). Instead, each model should be seen as accounting for various stages in the processing of information. Put differently, we can think of the bottleneck as supplying the information that gets divided among the pie. That is to say once information passes through the bottleneck, there are then limits to how much information

CHAPTER 6   ATTENTION (OR LACK THEREOF)

can be attended to simultaneously. Thus, the two models actually work in concert and really just describe different steps in the same process.[4]

The limits on attention are somewhat fixed (per each person) and are a byproduct, if not the direct consequence, of limits on what is called working memory (which we'll explore in more depth in Chapter 9).

Distractions: What Were We Talking About?
How Much Is Too Much: Cognitive Limitations

We are constantly being bombarded with information, regardless if screens are involved or not. Even when hiking in the woods, we are surrounded with sensory information that floods our senses. In this example, the information could include the tactile feeling of a soft breeze, the sound of the aforementioned breeze rustling leaves on overhead trees and the crinkling of those fallen and now underfoot, the views of the path ahead and the canopy overhead, and the smells of various flora. Even though in this fictional forest, we may think of it as "calm" or "peaceful," our brains are still gathering and processing massive amounts of information.

So why don't we consider the above example (i.e., nature) overwhelming?

It is because the information is all coming through different sensory inputs (e.g., our eyes, ears, etc.) rather than being forced through only one or two. Put in terms from the last section, there is more of our attentional pie to go around when we aren't solely focused on only one input stream (e.g., our eyes). To this end, studies have shown that when you overload one sensory unit, like the visual one, it depletes the resources that could be used for other cognitive processes. For example, on cable news, there is

---

[4] Angelopoulou, E., & Drigas, A. (2021). Working memory, attention and their relationship: A theoretical overview. *Research, Society and Development, 10*(5), e46410515288–e46410515288.

CHAPTER 6   ATTENTION (OR LACK THEREOF)

often a ticker that scrolls at the bottom of the screen (i.e., a chyron), which more often than not displays updates and information that is different from what newscasters and anchors are discussing. Indeed, when the news crawl (or ticker or chyron) is present, the amount of information that is *retained* is reduced by as much as 10%.[5]

I Hear What You're Seeing: Competing Inputs

With limited bandwidth to dedicate to information processing, our minds have to develop ways to prioritize which streams to pay attention to. With our discussion of the attentional "pie," we covered how there are only so many resources to go around; thus, our brains have to have the flexibility to move our focus from what is not important to what could be critical. One way to illustrate this situation is with the so-called "cocktail party effect."[6]

Imagine you're at a fairly crowded party. The canapes are circulating (oooh...pigs in a blanket!), and you've settled into one of the self-organizing conversational circles that always seem to form. You know the ones, little three-to-six-person groupings that congregate in various locations at the party to discuss similar topics or trade juicy gossip about people, places, and things.

---

[5] Bergen, L., Grimes, T., & Potter, D. (2005). How attention partitions itself during simultaneous message presentations. *Human Communication Research, 31*(3), 311–336.

[6] Wood, N. L., & Cowan, N. (1995). The cocktail party phenomenon revisited: attention and memory in the classic selective listening procedure of Cherry (1953). *Journal of Experimental Psychology: General, 124*(3), 243.

CHAPTER 6   ATTENTION (OR LACK THEREOF)

At this time, you happen to have settled on one such grouping in the kitchen around the island (a smart move – always be close to the source of food) and are having an in-depth conversation with several people about how you can't believe that [insert favorite singer here] has been rumored to be [insert gasp-worthy activity here] with [insert equally famous person here]. While you are thoroughly engrossed in this conversation (and why wouldn't you be) the din of the other conversational groupings becomes less and less distracting, allowing you to focus on the topic at hand.

Until you hear your name coming from one of those groups.

But how did you hear it? After all, nobody is yelling or screaming your name. They are saying it at the exact same level as the rest of their conversation, which up to now you had been successfully zoning out.

This ability to disregard irrelevant stimuli (conversations outside your little conclave) until it becomes relevant (hello – that was YOUR name that somebody said) shows that our minds are actually dedicating some

133

CHAPTER 6 ATTENTION (OR LACK THEREOF)

resources to monitoring what is going on in the background, even when we're not consciously aware of it. It also shows how important the notion of relevance is to grabbing our attention.

Just like with sensory adaptation (see Chapter 5), our cognitive systems are designed to respond to changes. In this instance, however, the change is related to whether or not we should care about the focus of other conversations and is a byproduct of what is called perceptual load. In perceptual load theory, there is a consistent monitoring that will redirect our focus when it is required to do so by the circumstances at hand.[7]

This ability to override distractions and cut through what is literally noise is a byproduct of many[8] things that we find relevant. Indeed, the fact that we have MANY relevant things in our lives can be exploited for design purposes.

For example, true customization of email messages that includes our first and last name has been shown to have positive results in fundraising campaigns, with increased rates of everything from the percentage of unsolicited emails that get opened to the number and quantity of donations. Ultimately, both of these situations, the cocktail party example and the email one, are about exploiting personal relevance (i.e., one's name) to heighten a signal relative to the noise (which of course is literal in the first example and figurative in the second vis-à-vis all of the unsolicited emails that one receives).

---

[7] Murphy, S., Spence, C., & Dalton, P. (2017). Auditory perceptual load: A review. *Hearing Research*, *352*, 40–48.

[8] Munz, K. P., Jung, M. H., & Alter, A. L. (2020). Name similarity encourages generosity: A field experiment in email personalization. *Marketing Science*, *39*(6), 1071–1091.

CHAPTER 6   ATTENTION (OR LACK THEREOF)

Business As Usual: Habits and Routines

Nothing so needs reforming as other people's habits.

—Mark Twain

"That's a bad habit you got there." "It's just routine – business as usual."

Habits and routines are ways in which we help more efficiently manage our attention problems. We have a habit (see what I did there) of confusing the words habit and routine, but they are actually two distinct ideas. Let's take a moment to talk about the difference.

A **habit** is an automatic, often subconscious, behavior triggered by specific cues. Stopping at a red light, biting your nails when nervous, and reaching for your phone first thing in the morning are examples of habits. They are triggered by the time of day, a feeling, or another stimulus and performed without thinking.

A **routine** is a consciously planned sequence of actions that requires effort and intention. A morning routine that involves waking up, exercising, showering, and then eating breakfast is a routine. It's a series of actions that you plan and execute each day, usually in a specific order. You may feel uncomfortable or "off" if you miss your routine – routines are excellent ways to create a context or environment to promote a particular activity. For example, you might have a night time routine of

- Take a hot shower

- Stretch

- Read a chapter of a book that prepares your mind for bed and makes going to sleep easier

Or you might have a routine of

- Making a cup of tea

- Selecting a small cookie

135

CHAPTER 6 ATTENTION (OR LACK THEREOF)

- Sitting in your armchair that prepares you for writing a chapter in a book (hmmm...wonder where this example came from?)

Whatever your routine, the series of actions are *thoughtful* and *deliberate* rather than automatic.

AND THEN THERE'S AUTOMATICITY

Just to keep it fun, there's also automaticity — the formation of knowledge that is so ingrained as to not require conscious effort to repeat. The multiplication tables you learned in elementary school are a great example of this — I can ask what's 2×2 and chances are good you have the answer before you've finished reading the question. Automaticity lets us remember our phone numbers and addresses, the driving route home, and the words to "Let It Go" from *Frozen* without thinking about it. Which is great when you're rocking it at karaoke, but not so great when you're tired and driving home at night after a party. Sometimes, you have to interrupt the automaticity to pay attention to what's happening when something occurs that may not be part of the ordinary — like that deer that just ran in front of your car.

Habits tend to be more ingrained and difficult to change, whereas routines are more flexible and can be adjusted based on circumstances or goals. Both play important roles in structuring our daily lives, but they operate differently in terms of consciousness, effort, and adaptability. You'll notice the key difference here is conscious effort. A habit is something that you do without thinking. A routine is something you think about and plan but do repeatedly. Both routines and habits can help reduce the conscious effort needed to do a thing either by making it more automated, freeing up your limited attentional resources (getting you more of that pie – mmmmmm...pie...), or preparing your mind to tackle a specific type of task (sleeping, writing, going to work, etc.) which in turn makes it easier to take on that task.

136

CHAPTER 6   ATTENTION (OR LACK THEREOF)

Designing for Routines

Designing for routines can provide significant benefits both for users and for the success of your product. By aligning your designs with natural human behaviors, you can offer structure, consistency, and support in a way that can significantly improve users' experience. But how does one design to support routines?

We've harped on this from the beginning, and this won't be the last time you hear it – understand your user. Before you do anything else – before you define your problem, come up with a solution, design a logo, or answer that CEO's email – take the time to really understand your user. The underlying secret to creating good designs is understanding who you are designing for, and that is doubly true when creating products that support routines. If you are creating a product that helps people organize their mornings and you don't make space for "make the coffee," you're automatically alienating all of the coffee drinkers (and there are a lot of us) out there.

But what does it mean to understand your user? We're glad you asked. While we'll beat this drum a lot ("understand your user" **BANG**; "you are not your user" **BANG**), it's no small feat to get this information. There is no easy or fast way to understand your user. You have to (say it with us): DO THE RESEARCH.

What does that look like?

1. Identify what type of people you want to target with your product. Are they Boomers? Karens? Dog lovers? This is the key area you want to concentrate your research efforts on. Creating a dog walking app, for example, will automatically narrow your group of interest to people who have dogs that need to be walked and people who walk dogs.

137

CHAPTER 6 ATTENTION (OR LACK THEREOF)

2. Survey the group you are interested in, specifically around the behaviors you want to support. For our dog walking app, we might be interested in the number of times you need your dog walked every day/week/month and at what time of day. We could ask questions about what potential consumers believe makes a good potential dog walker, how they typically communicate with dog walkers today, what the impact of cancellations is on them, and how they handle it. We use this survey to build a picture of our target user called a persona (we'll talk about this more in Chapter 7) and a picture of how they interact with our product called a journey map.

3. Refine your persona and journey map by conducting interviews. Interviews will help you validate your artifacts but also fill in the gaps (the whys) that a survey can never get to.

PROTIP!

Understand what you are going to do with the data before you gather it. In this way, you focus your questions and ensure you are using your and your audiences' time effectively. For our dog walker app, we wouldn't want to ask about their morning personal hygiene routine because we wouldn't be able to DO anything with that information. In any research effort, one of the best ways to create efficiency and cut down on wasted data collection is to know what you will do differently once you have the information – what decision will your data inform.

CHAPTER 6 ATTENTION (OR LACK THEREOF)

With your data in hand, it's time to start looking at designs to support or help create routines. Typical designs that help support routines include

- Reminders
- Calendars
- Repeatable checklists

Each of these can be created in a way to support the specific routine you want to reinforce. For example, consider our dog walking app, Floofer's Walk. We know from our journey map that our typical user has a routine in which they need or want their dog walked on the same days at the same time every week. A key feature to design would be a repeating calendar entry to reserve service for that same time slot weekly. The data tells us how to design the most valuable but least complicated calendar reservation system (need to be able to make something repeat both within the day and weekly but not monthly or yearly) that will serve the greatest number of people. This supports the routine they have (or want to have) around dog walking, making the act of reserving a dog walker through Floofer's Walk a walk in the park.

CHAPTER 6   ATTENTION (OR LACK THEREOF)

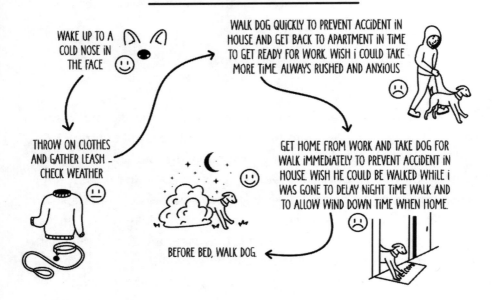

Designing for Habits

Now, we've already pointed out that habits are automatic – something you do based on a trigger. So how do we design for that? The first step is to understand the parts of what makes a habit. This is called the habit loop.[9]

---

[9] Duhigg, 2012. Power of Habit.

CHAPTER 6   ATTENTION (OR LACK THEREOF)

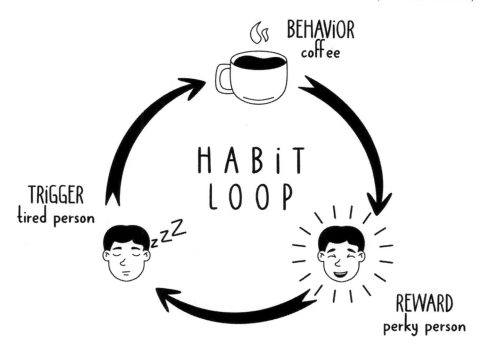

The habit loop consists of three parts: a cue/trigger, a routine/behavior, and a reward. The cue is what triggers the habit. It can be anything (or anyone) – a smell, a situation, a feeling, a sound. It is the stimulus that elicits the behavior we think of as the habit. The routine is the behavior itself – eating, surfing the web, chewing your lip. And the reward is what makes us want to repeat the cycle – the pleasure of tasting sugar, the satisfaction of finishing a book, the endorphins from a run. Each of these steps is necessary in forming and maintaining habits. It's interesting to break down habits this way because it lets us see how we are ceding control and how (if we need to) we can reexert that control to change, add, or eliminate habits to enable other things we'd rather be spending our limited attention on. By changing, pairing, or removing a trigger, or by altering the reward, we can disrupt the habit and change its outcome.

CHAPTER 6   ATTENTION (OR LACK THEREOF)

Hooked on Habits

Nir Eyal's book, *Hooked: How to Build Habit-Forming Products*, introduces a variant on the habit loop called the hook model that helps envision how we, as designers, can help create and enforce habits.

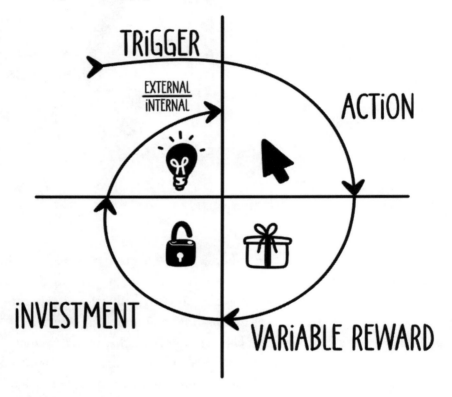

The hook model has four steps rather than three: trigger, action, variable reward, and investment. The trigger and action are the same as the habit loop – a stimulus and a response. The reward is changed to variable reward – that is, the reward is not always the same. This variability keeps our brains stimulated with the unexpected/novelty of things. Investment is getting the user to put something of value (time, effort, data, or social capital) into the product, increasing commitment and making future use more likely.

CHAPTER 6 ATTENTION (OR LACK THEREOF)

Let's take a look at how that looks when designing.

Triggers

External triggers are the easiest to create – a badge on your application telling you there's an email or a new Instagram post to look at, an alarm that tells you it's time for your meeting, a flashing light to tell you to pull over to the side of the road and let the fire engine go past. External triggers are a sensory indicator (visual, auditory, or tactile, usually) that notify the user that an action is required or desired. Over time, external triggers can cultivate internal triggers, where users start associating certain feelings like boredom or stress with going to a product like email.

Variable Reinforcement

Incorporating different types of rewards helps us tap into user motivations, like social validation (likes or comments), rewards of the hunt (discovering new content), or rewards of self-improvement (tracking progress). For example, Facebook uses variable rewards in the feed, showing fresh news, friend updates, and posts, which makes us curious about what we might see next time we log in. Variable reinforcements have been shown to be far more effective than fixed reinforcement (i.e., always present) in changing and reinforcing behaviors.[10] For designers, there are many ways we can provide variable reinforcement.

- **Levels**: Tiered systems that let users advance to higher levels as they gain points or complete tasks offer a sense of progression and growth.

---

[10] Skinner, B. F. (2014). *Contingencies of reinforcement: A theoretical analysis* (Vol. 3). BF Skinner Foundation.

CHAPTER 6   ATTENTION (OR LACK THEREOF)

- **Progress Indicators**: "Your profile is 50% complete" "You have finished 90% of this survey" – progress indicators are reminders to our users that let them know how far along they are toward a particular goal. Seeing this progress encourages people to complete the task to achieve that 100%.

- **Random Rewards**: Unexpected rewards, information, or interactions engage a user's curiosity, encouraging them to extend their interaction to see what happens next. And the seeming "randomness" of them plays right into the effectiveness of a variable reinforcement schedule.

The important aspect of variable reinforcement in all these techniques is, well, the variability. Levels are only effective if the reward changes as the level changes. Progress indicators are useful when you have something with distinct steps you can take to move the needle. And random rewards need to be random (duh!) but also interesting, delightful, surprising, and relevant.

Dr. Baker once read a story in which a young boy was asked by his mother what he had in his pocket. "Confetti" was his prompt answer. His mother, nonplussed, asked why on earth he would have confetti in his pocket. "In case of emergency!", he told her. She loved this idea of emergency confetti so much, she started carrying confetti in her jacket pockets wherever she went (Dr. Rekart can attest to the veracity of this statement – he once found a bag of confetti in his home after she'd visited).

Someone on her team would do something amazing, and she would throw confetti in celebration. It was so unexpected but also so delightful that the person who was the subject of the confetti would inevitably laugh with happy surprise. She made sure not to throw confetti ALL the time but rather tried to be more random about it, only sometimes breaking it out to celebrate. Her coworkers loved it and started trying to figure out what they could do to get her to throw the confetti. Providing a delightful experience that reinforces positive behavior can be as simple as a pocketful of emergency confetti.

144

CHAPTER 6 ATTENTION (OR LACK THEREOF)

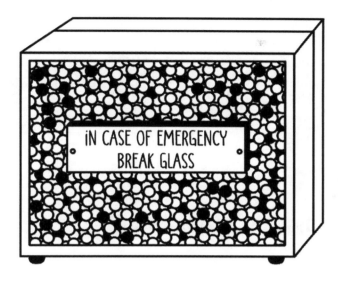

Investment

Getting investment from your users is the key to turning the habit loop into a hook. When someone is invested in your workflow (remember sunk-cost bias?), they are far more likely to come back again and again. It could be an investment of time and effort (time spent customizing an avatar, creating favorites), an investment of data (saving your credit card information for you, saving your past purchases), or an investment in social capital (creating friend groups, answering questions). Investment asks the user to bring something to the table and share it which triggers the endowment effect (the fact that we value something more when we feel ownership of it).

In an experiment done by Kahneman, Knetsch, and Thaler (2009), subjects were given a coffee mug or a bar of chocolate, then offered to trade it for whichever item they did not have. Regardless of what they

CHAPTER 6   ATTENTION (OR LACK THEREOF)

originally had been given, they were reluctant to give up what they had, perceiving it to be more valuable than what was offered in the trade. This result, attributed to the endowment effect,[11] is what drives us to cling to that hole-ridden old T-shirt, to try to sell our old couch instead of donating it, and what makes us spend more time on social media even though we complain about how much time we spend on social media. We become invested in these things which clouds our ability to logically determine the value of them.

The Mythical Multitasking Beast

With the constant interruptions of modern life, are we doomed to a world of distractibility? Are our children, the new "digital natives," born with a proclivity for multi-tasking due to the constant influx of information and stimulation? The answer is largely: no, but read on to learn why.

Old Problem; New Wrapper

In 2015, a report from Microsoft caught the attention (couldn't resist) of the media. This report suggested that members of Gen Z (those born between the years of 1997 and 2012)[12] had attention spans that were only eight seconds long. This "study" went viral quickly because it seemed to confirm the stereotype that "kids today" just couldn't stop staring at their phones and were doomed to be less attentive than even goldfish (yep, that was a

---

[11] Experimental Tests of the Endowment Effect and the Coase Theorem," Journal of Political Economy 98, 1325–1348.

[12] Although it should be noted that the years that demarcate different generations are not based in hard science but are contrivances that make it easier to think about groups. Because of this the calculation of when a generation begins and ends is subject to different logic and formulae, depending on the organization. https://www.pewresearch.org/short-reads/2019/01/17/where-millennials-end-and-generation-z-begins/.

CHAPTER 6   ATTENTION (OR LACK THEREOF)

comparison that was made, with goldfish touted as having attention spans that were 12.5% longer than those of Gen Z). Even more concerning than being less attentive than a goldfish was that this study claimed that this was a sizable drop of attention spans from those of Millennials (born between 1981 and 1996), who reportedly had an average attention span of only 12 seconds. This study thus spawned numerous articles and blog posts on how to reach the ever-moving target of young eyeballs, which in turn, we have no doubt, resulted in meeting after meeting about how to reach young consumers and how to better meet the needs of junior colleagues and coworkers.

Except none of it was true.[13]

Turns out that the "data" showing the "eight-second span" weren't even collected by Microsoft but from two studies conducted by an outside firm. The first included a small set of Canadian users and wasn't subjected to peer review or any other standards. And the second may have been fabricated entirely![14]

Though the BBC investigation disproving the validity of these findings was published in 2017, a Google search in 2024 still returns all of the articles written for designers, developers, educators, and industry leaders about how to compensate for the woeful state of youth today (despite the fact that Microsoft has removed the study from the web). Because these findings align with what individuals from older generations *feel* is true, the eight-second myth persists.

Indeed, the lamentations of one generation about the inattention and distractibility of youth is nothing new. In this famous passage from a series of correspondences from a father to his son, the father, Phil Stanhope, wrote:

---

[13] https://www.bbc.com/news/health-38896790

[14] https://www.forbes.com/sites/shanesnow/2023/01/16/science-shows-humans-have-massive-capacity-for-sustained-attention-and-storytelling-unlocks-it/?sh=729b67281a38; https://www.fastcompany.com/91023619/8-second-attention-span-is-bs-this-is-why

CHAPTER 6 ATTENTION (OR LACK THEREOF)

*This steady and undissipated attention to one object
is a sure mark of a superior genius; as hurry, bustle,
and agitation are the never-failing symptoms of a
weak and frivolous mind.*[15]

You may have noticed that the prose is a bit formal for a simple note from a dad to his kid, but that may reflect any of the following facts about the originator of this nugget of wisdom:

- The author's full name wasn't "Phil" but was actually Philip Dormer Stanhope, the fourth Earl of Chesterfield.

- He was an accomplished, serious man, with positions and titles ranging from being a member of the British House of Lords to Ambassador to the Hague to the "Lord-Lieutenant" of Ireland.[16]

- And, most importantly, he wrote the letter on April 14, 1747.

Thus, whether or not one is particularly prone to distraction is likely more a byproduct of how more mature individuals view youth than a de-evolutionary backslide of human cognition due to technological innovations (or put differently tech is NOT making us dumb).

---

[15] https://www.gutenberg.org/files/3361/3361-h/3361-h.htm
[16] https://www.english-heritage.org.uk/visit/places/rangers-house-the-wernher-collection/history-and-stories/lord-chesterfield/#:~:text=The%2018th%2Dcentury%20politician%20Philip,Letters%20to%20his%20illegitimate%20son

CHAPTER 6   ATTENTION (OR LACK THEREOF)

Not surprisingly, laboratory studies easily dismiss the notion that somehow attention spans are shrinking, with demonstrations of comparable levels of attention for young and older adults,[17] with young adults having slightly *longer* attention spans than older ones.

Indeed, even outside of the laboratory, there are consumer data that reinforce that there is nothing irredeemable or irreconcilable about Gen Z's struggles with attention, with recent increases in physical book sales and attention to so-called "long form" videos as proof-positive of attention spans that are absolutely fine.[18]

Now, how does one gain the title of "Lord-Lieutenant"?

Whither Hast Thou Gone the Yeti, Unicorn, and Multitasker?

Now that we have dispelled the myth that "kids today" have attention spans that goldfish would scoff at (can one scoff underwater?), we can address the next attentional fallacy regarding so-called "digital natives": their ability to multitask. Almost 20 years ago, *Time Magazine* lamented the fact that children who grew up with modern technology, alternatively referred to as "Millennials" or "Gen Y," were using digital devices with alarming regularity (i.e., media multitasking).[19]

---

[17] Simon, A. J., Gallen, C. L., Ziegler, D. A., Mishra, J., Marco, E. J., Anguera, J. A., & Gazzaley, A. (2023). Quantifying attention span across the lifespan. *Frontiers in Cognition, 2*, 1207428.

[18] https://www.mckinsey.com/~/media/mckinsey/email/genz/2022/11/29/2022-11-29b.html

[19] https://content.time.com/time/classroom/glenfall2006/pdfs/the_multitasking_generation.pdf

CHAPTER 6   ATTENTION (OR LACK THEREOF)

Multitasking is defined as performing two or more activities simultaneously. When those activities tap into different cognitive resources or are automatic, then certainly multitasking can be achieved, like holding a conversation while walking with a friend or listening to music while driving a car. When both activities require the same or overlapping neural processes however, like reading and, well, really anything else, then multitasking is in fact almost an impossibility.

Despite the cognitive constraints that our brains place on us, there continues to be a misplaced belief that somehow kids who have been raised with digital devices, like those of the Millennial, Gen Z, and later generations, are more adept at using them while engaging in other activities, like learning, reading, or making decisions. In fact, the data continue to show that the opposite is true.

CHAPTER 6   ATTENTION (OR LACK THEREOF)

First, studies have shown that most individuals have an overinflated sense of how good they are at multitasking,[20] particularly when it comes to "media multitasking," which is the usage of multiple devices that require attention, such as a laptop and a cell phone. And it gets worse. Those individuals who feel that they are the best at multitasking are in actuality the worst at juggling multiple things simultaneously.

Second, our brains have not adapted to be able to make sense of multiple visual or auditory streams of information simultaneously; thus, no one generation is better suited to multitask than another. The cognitive limitations of the brains of Baby Boomers and their children, the GenXers, are no different than those of later generations[21].

And finally, multitasking behaviors have been shown to be correlated with impairments in cognitive functioning, not the other way around. Cain and colleagues (2016[22]) showed that those adolescents (as evaluated in 2015) who spent more hours of the day media multitasking also had lower scores on working memory and reduced performance in academic tests of English and mathematical ability. Because these data are correlational, we cannot know whether individuals with reduced working memory are more attracted to using multiple devices or whether using multiple devices had an impact on their scores, but we do know that there is no inherent advantage held by users – whether young or not – who use tech all of the time.

Taken together, these results show that one shouldn't make design decisions that will tax the amount of attention required to either understand a feature or site, regardless of the anticipated age of the users.

---

[20] Sanbonmatsu, D. M., Strayer, D. L., Medeiros-Ward, N., & Watson, J. M. (2013). Who multi-tasks and why? Multi-tasking ability, perceived multi-tasking ability, impulsivity, and sensation seeking. *PloS one, 8*(1), e54402.

[21] Gawda, B., & Korniluk, A. (2022). Multitasking among modern digital generations Y and Z. *Journal of Modern Science, 49*(2), 421–430.

[22] Cain, M. S., Leonard, J. A., Gabrieli, J. D., & Finn, A. S. (2016). Media multitasking in adolescence. *Psychonomic bulletin & review, 23*, 1932–1941.

CHAPTER 6   ATTENTION (OR LACK THEREOF)

Context Switching

So, if multitasking is a unicorn (i.e., completely fictional), what ARE those crazy digital natives doing that makes them think they are multitasking? They are context switching. In computers, context switching happens when one process gives way to another because of a lack of CPU to complete the second process. Actually, it's the same for humans. When you stop writing your report because you get an email which you start to respond to but then stop because you got a Teams message from your boss asking you about a project that you start to answer but then stop because someone walked up to your desk to ask where you were on that report you were writing – you are experiencing context switching. It can feel like you are getting all of the things – the report, the email, the question – done and that you are being oh so efficient. But this is one of those cases where your perception of your productivity is fooling you. Research[23] has shown that it takes, on average, nine and a half minutes for you to move from one task to the next effectively. Think about that a moment – for every time you switch tasks, it will be almost ten minutes before you are really paying attention to the next thing. And, worse than that, context switching wears us out. Each time you have to get into a new thing (even when you are returning to the old thing that is now the new thing), your cognitive load increases.

Designing for Attention: Make Hammers Not Heroin

Which brings us back to design choices. We've talked about how to "hook" our users on our products/workflows. But is that always a good thing? For a classic example of good intentions gone bad, let's take a look at email.

---

[23] https://assets.qatalog.com/language.work/qatalog-2021-workgeist-report.pdf

CHAPTER 6 ATTENTION (OR LACK THEREOF)

These days, email is ubiquitous and omnipresent. Originally designed as an asynchronous communication tool, modeled after our postal or "snail" mail, email offered a way to send a message quickly and efficiently in a way that the recipient could respond to it when they had a moment. Imagine the use case construction – designers likely thought it would function like taking a walk to the mailbox, something done once a day, to then pursue the message therein at a leisurely pace, deleting those that weren't interesting or relevant, planning to deal with those that needed action at a future date, and setting aside time to reply to those friendly missives from friends and family. How far we have fallen from that skeuomorphic ideal!

Today's emails are treated as time-sensitive demands, documentation systems, and accountability trackers. Design elements such as badging (the little red dot that tells you that you have unread messages) have supported this assertion, providing an unspoken expectation that you will want to check your email every time a new message comes in. Send receipts tell the sender whether you have received their email. And advanced search capabilities lets you send pages of documentation about a process/procedure/encounter in email (with folders so you can bask in the illusion of organization).

While all of that sounds like a condemnation of the tool itself, that is not the case. It's not email that is objectionable; it's how we have designed email to take advantage of the average person. Remember our design elements to hook someone on our product? Email has them all:

- Triggers (ding, you've got mail, see the little red badge?)

- Variable reinforcement (what's new in my inbox? Look how many unread messages I have!)

- Investment (I've saved those emails about the contract in this folder, I have everyone's email addresses in my contacts.)

153

CHAPTER 6   ATTENTION (OR LACK THEREOF)

And while this is effective in keeping us in email, the question is – should we? Are we serving our users in the best possible way by constantly pulling them into a habit loop of checking email every time there's something new or they're bored?

We suggest the answer is "no." And this comes back to experience with any number of C-suite executives who want the new interface to be "gamified," "sexy," or "addictive" and who measure success by how frequently users access their software.

Just. Don't.

Software, and in particular business software, is not meant to be addictive or sexy or anything other than useful. And while being attractive does, in fact, make people feel like something is more usable,[24] success is not measured in how many times people come to your application but in whether they are able to achieve their goals with your software quickly and efficiently so they can get back to the rest of their lives (and finish that *Legend of Zelda* level). Business software is about making hammers – a hammer is good for a particular task, and it is really really good at it. If you found yourself wanting to use a hammer every day even though you don't have anything you need to nail down, you might question your sanity (as well you should). But we don't afford software the same type of expectation or examination. Modern software should be designed with the user's *success* in mind. That requires a careful examination of the tasks the user needs to do, what the software can do to make that task easier, and a willingness to support a single-threaded experience especially in a multiproduct platform experience.

---

[24] Sonderegger, A., & Sauer, J. (2010). The influence of design aesthetics in usability testing: Effects on user performance and perceived usability. *Applied ergonomics, 41*(3), 403–410.

CHAPTER 6   ATTENTION (OR LACK THEREOF)

Flow

One step toward making your users successful is providing a seamless experience. There are times when we not only become engrossed in a demanding activity, like playing a sport or an instrument or (successfully) working through complicated formulae or coding schemes, but appear to be doing so effortlessly. This almost autopilot like state is referred to as "flow" and has been the subject of fair amount of research since first characterized by the Hungarian-American researcher Mihaly Csikszentmihalyi about 50 years ago.[25]

Other than being synonymous with an individual who had a seemingly unpronounceable surname ("Csikszentmihalyi" is pronounced "Cheek – sent – me – hi"), for positive psychologists, flow represents a pinnacle of human experience. For someone to enter into a flow state, two criteria must be met. First, the individual must be operating at the top of their skill level for a particular task. Second, the situation must be one that is sufficiently challenging as to push the limits and test the skill of the individual. This relationship is depicted nicely with a simple quadrant system that plots personal skill and situational challenge on the X and Y axes, respectively.[26]

---

[25] Csikszentmihalyi, M., & Larson, R. (2014). *Flow and the foundations of positive psychology* (Vol. 10, pp. 978–94). Dordrecht: Springer.

[26] Adapted from Abuhamdeh, S. (2020). Investigating the "flow" experience: Key conceptual and operational issues. *Frontiers in psychology, 11,* 158.

CHAPTER 6   ATTENTION (OR LACK THEREOF)

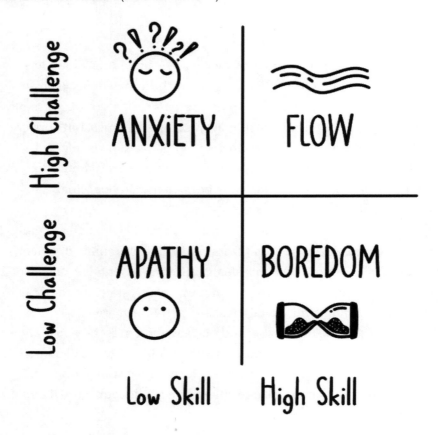

Although everyday interactions with design may not be sufficient to evoke flow for a large number of users (and certainly not at scale), it is important to note that tasks that are overly challenging but are used by individuals with low to no skill will be anxiety-provoking (think requiring the average web user to code in HTML in order to clear their cart). More likely, however, the utility of research on this phenomenon is more applicable to use cases that the designer herself will encounter as she works on projects in her portfolio. For although the critical parameters for entering into a flow state are challenge and skill, once immersed, Csikszentmihalyi, among others, describe the feeling of being *in the flow* as one of pleasure and enjoyment.

CHAPTER 6 ATTENTION (OR LACK THEREOF)

Recap

- Attention is limited – our brains take in huge amounts of data but have to filter out most of it to be able to function.

- Habits and routines are different. Both require a deep understanding of the user to create effective designs.

- A variation on the habit loop that includes variable reinforcement and investment can be used to create and reinforce habitual interactions for users.

- Young adults from more recent generational cohorts (e.g., Gen Z, Millennials) do not have shorter attention spans than adults from later ones (e.g., Gen X, Baby Boomers).

- Multitasking is a lying liar that lies.

- Context switching is how we fool ourselves into thinking we are multitasking, and it's wearing us out.

- Designing compelling habits that induce us to multitask is tempting and ignores the needs of the user to have fewer interruptions. Designing with attentional limitations in mind leads to more long-lasting designs with happier users.

- Flow is an immersive cognitive state that provides a relatively distraction-free zone of productivity for highly skilled individuals in their craft.

157

CHAPTER 6   ATTENTION (OR LACK THEREOF)

Before You Go…

Though we've dedicated a good deal of this chapter to discussing how important attention and engagement are, we would be remiss if we didn't acknowledge that our brains are also quite adept at so-called "offline processing." What we mean by this is the fact that complex problems or challenges continue to be examined in our minds, sometimes to the point of being solved when we are least expecting them to be. This phenomenon was exploited to great effect by Thomas Edison, yes *that* Thomas Edison, the one with over 1000 patents, who famously believed that four hours of sleep a night was more than enough.[27] Supposedly, Edison would take naps throughout the day with two steel balls clenched in each hand, so that when he began to drift off the balls would drop and the resulting clatter would awaken him. The rationale for being awoken wasn't to maximize productivity, however, but to take advantage of what has come to be called hypnagogic problem-solving, which is a highly creative state that we can enter into just as we're beginning to drift off to sleep (similarly, the stereotypical chin droop and subsequent spasm that takes place when folks are drifting off while sitting up is referred to as a "hypnagogic nod"). This phenomenon, which has been replicated in laboratory settings, only works if a few criteria are met. First, you must have been thinking about the problem or challenge while awake. Second, you need to begin thinking about the issue as you relax. Third, and most importantly, you need to awaken before you fall into true sleep[28] (in the laboratory re-creation, a

---

[27] Sherard, R.H. (1889). With Mr. Edison on the Eiffel Tower, *Scientific American, 61*(11), 166.

[28] Lacaux, C., Andrillon, T., Bastoul, C., Idir, Y., Fonteix-Galet, A., Arnulf, I., & Oudiette, D. (2021). Sleep onset is a creative sweet spot. *Science Advances, 7*(50), eabj5866.

CHAPTER 6   ATTENTION (OR LACK THEREOF)

glass was used instead of steel balls, but to the same effect). Though the reason why our brain seems particularly primed to solve problems in this hypnagogic state is still being explored, it may be worth trying, unlike some of Edison's other habits (like smoking over 20 cigars a day).

CHAPTER 7

The Evolution and Revolution of People

RB: It's so wild how our vibe switches up as we glow up. There are, like, a ton of different ways our needs shift, based on what we know, where we're from, and even if we're a Boomer or a Zoomer.

JLR: Hang on, I think I need a translator…

RB: Exactly! Linguistics is a simple differentiator when it comes to culture, age, and expertise. But there are so many other things as well – physical needs/accommodations, cultural biases, and domain knowledge. It's key to take these things into account when we're designing solutions to make sure our users get a main character vibe. IYKYK.

JLR: Groovy. It's such a bummer when some dude designs something gnarly for a square and you have to rap with support just to dig it.

RB: Like, totally! I, like, want legit products that get me. And that means, like, taking time to understand your audience to make, like, a wicked product so they don't, like, bounce.

JLR: That's lit!

RB: No cap, for real.

© Jerome L. Rekart and Rebecca Baker 2025
J. L. Rekart and R. Baker, *Designing for Human Intelligence in an Artificial Intelligence World*,
https://doi.org/10.1007/979-8-8688-1418-1_7

CHAPTER 7   THE EVOLUTION AND REVOLUTION OF PEOPLE

Back in My Day: Age-Related Design Considerations

When designing for the masses, one must consider how users from various cohorts will interact with a given product. One of the easiest demographic variables to consider – in particular, because it happens to all of us (or at least those still able to read this) – is aging. Perhaps not surprisingly given other research shared thus far in this volume, there is no unequivocal set of findings as they relate to *biological* underpinnings that may explain any cognitive and perceptual characteristics of the aged. Indeed, even the term "aged" must be operationalized and qualified – aged relative to whom? Though when you were a teenager you may have thought about longevity vis-à-vis your own age (i.e., everyone older is ANCIENT), with some more trips around the sun (as with the two of us), you may start to think more in geological terms (and thus consider how incredibly NEW you are).

Even once terms like aged and elderly are properly qualified, whether or not there are cognitive impacts of accumulating decades requires even further refinement. This is because there is a great deal of variability among individuals, with some maintaining their full "capacities" well into their eighties, while many begin to see a gradual decline from the third decade into the sixth, with a steep decrement thereafter.[1] Given all of this, what research has shown somewhat generally is that tasks requiring perceptual speed do seem to decrease in the vast majority of individuals.[2] This means that if you know that your site will be visited by a large cross-section of

---

[1] Sánchez-Izquierdo, M., & Fernández-Ballesteros, R. (2021). Cognition in healthy aging. *International Journal of Environmental Research and Public Health*, *18*(3), 962.

[2] Savage, R. D., Britton, P. G., Bolton, N., & Hall, E. H. (2024). *Intellectual functioning in the aged*. Taylor & Francis.

CHAPTER 7 THE EVOLUTION AND REVOLUTION OF PEOPLE

individuals or even targets those old enough to remember when there were two Germanys,[3] then you will want to ensure that time-bound tasks are minimized, if used at all.

When we think about age as representing a cultural element, rather than just a cognitive one (as discussed above), generational groups (or generations) are often used as a shorthand to identify individuals of roughly the same age (within a 15–20-year span) who have experienced similar events and phenomena. For most generations, the years that delineate when one generation ends and another begins are not well defined. The only exception to this is the agreed upon 19-year span of the "Baby Boom," which is broadly used as it was defined by the U.S. Census Bureau. Though there is no other "official" guide that determines what birth years gain someone entry into particular generations, most outlets and researchers follow the guidance of the Pew Foundation (for historical reasons) to draw rough, chronological boundaries.[4]

Thus, the following are reported as the beginning and ending dates (and the corresponding label) for each generational cohort. To determine which cohort you belong, just locate the interval containing your birth year:

- 1928–1945: Silent Generation

- 1946–1964: Baby Boomers

- 1965–1980: Generation X (this one is the COOLEST...no, we're not biased AT ALL)

---

[3] If you don't know what "two Germanys" means, see: https://www.archives.gov/exhibits/eyewitness/html.php?section=10.

[4] https://www.pewresearch.org/short-reads/2019/01/17/where-millennials-end-and-generation-z-begins/

163

CHAPTER 7   THE EVOLUTION AND REVOLUTION OF PEOPLE

- 1981–1996: Millennials

- 1997–2012: Generation Z

- 2013–present: Post-Generation Z (at the time of this writing, Pew has not officially labeled this group,[5] though they have received various labels, including "Generation Alpha" and "Generation Glass," the latter of which refers to the fact that members of this generation have never lived in a world without smartphones, tablets, or other web-enabled screens)

The usefulness of generational cohorts is that they *can* provide some means for categorizing individuals who may have experienced different global trends or phenomena at roughly the same time, which can influence perspectives, decisions, and cultural considerations. For example, most members of Generation Z spent at least some part of their childhood or early adulthood living through a global pandemic (COVID-19), which may have lasting impacts on how members of this cohort generally view things. For example, individuals from this particular group may have similar lowered expectations for the future and have a similar acceptance for how "normal" it is to have your groceries delivered to your front porch.

One (of many) drawbacks to using generations as a lens for design, however, is that they become bait for stereotypes. Generational labels make it far too easy to assign behaviors, beliefs, and attitudes to large groups of individuals without considering that there is a great deal of within-group variability. Put differently, all things being equal, members of a given generation can vary as much (if not more) from one another (e.g., Boomers vs. Boomers) as they differ from members of other generations (e.g., Boomers vs. Millennials). Furthermore, it doesn't make sense to think

---

[5] https://www.pewresearch.org/short-reads/2023/05/22/how-pew-research-center-will-report-on-generations-moving-forward/

164

CHAPTER 7 THE EVOLUTION AND REVOLUTION OF PEOPLE

that a member of Generation X who was born in 1980 would somehow be *meaningfully* different from a Millennial born in 1981. Indeed, in the latter example, the Gen Xer would most likely have been a fourth grader in 1998 when the Millennial was a fellow elementary school denizen, only one year back in the third grade. These are some of the reasons why most psychologists and sociologists don't utilize these labels very often.[6] In fact, Pew itself is even starting to move away from such labels (see Ref. 4).

IYKYK: Expertise-Related Considerations

Another important consideration in design is domain expertise. This aspect can be especially difficult for designers because it is in our nature to design for beginners (a.k.a. ourselves). But, as we have said *ad nauseum* "You are not your user" and failing to design for the correct level of domain knowledge will not only alienate your users, it may cause your product to be difficult to use.

Yes, that is right – we just wrote that making something easy for beginners makes it difficult for *experts* to use. Consider a nurse using diagnostic software to enter prescriptions for multiple patients. They have to get the information into the system quickly and accurately, and they only have 10 minutes to enter the last 15 patients' worth of data. What if we designed the interface for them in such a way that

---

[6] Rudolph, C. W., Rauvola, R. S., Costanza, D. P., & Zacher, H. (2021). Generations and generational differences: Debunking myths in organizational science and practice and paving new paths forward. *Journal of business and psychology, 36,* 945–967.

CHAPTER 7   THE EVOLUTION AND REVOLUTION OF PEOPLE

- Every drug was spelled out
- All the codes that hospitals use for patient conditions were listed
- Each entry required a validation step in which you acknowledged that you had reviewed the patient's condition with the physician prior to obtaining this prescription order

This scenario would likely be a great experience for a beginner who is unfamiliar with the hospital lingo, the prescription order process, and the hospital protocols. However, for the nurse who has been working for the last six months at this hospital, this is a huge waste of time (time they don't have) and an increased opportunity for error.

Every domain has its own unique set of knowledge like this that can be used to accelerate and support the work for our users. But, you might be wondering, what about the beginners? It's important to understand what

CHAPTER 7   THE EVOLUTION AND REVOLUTION OF PEOPLE

portion of the population you are targeting are indeed beginners and how many of those will be using your product. Often the answer is a surprising "none" as beginners are still learning more basic tasks than your product covers. In addition, once your system gets implemented, training how to use it will likely become part of any orientation process. Regardless, it is still a good practice to provide "scaffolding" for your designs to help ease new users into your experience.

Scaffolding is a tried-and-true technique in education and gaming but is less well known when it comes to design. The idea behind scaffolding is to provide basic structures and knowledge for your audience that fall away or adapt as they grow in knowledge about the topic. By providing the *Goldilocks* level (not too much and not too little) of instruction or help, we achieve the Zone of Proximal Development[7] – a state of knowledge where someone is learning with just enough help so that they can learn what they need to do (system or human). Hitting the Zone of Proximal Development (yes, we know – it DOES sound like a TikTok diet fad or a mid-level club with mid DJs) is critical because if something is too easy, it may not be cataloged by the brain as being important enough to learn, and if it is too hard, it can induce anxiety (which is counterproductive to learning).

Of course achieving this is easier said than done. With variable audiences and potentially little real-time feedback (or the ability to respond to real-time feedback), it might seem difficult or even impossible to design a system to support multiple levels of domain expertise. Unsurprisingly, many systems index on either end of the spectrum (the beginner OR the expert) making the resulting experience either miserably difficult for the newbie or frustratingly time-consuming for the veteran. However, with a little forethought, we can provide designs that begin to address this issue. Here are three ways to begin:

---

[7] Vygotsky, L. S. (1978). Mind in society: The development of higher psychological processes. Cambridge, MA: Harvard University Press.

167

CHAPTER 7   THE EVOLUTION AND REVOLUTION OF PEOPLE

1. Break down the task at hand into small pieces. By deconstructing tasks into bite-size chunks, you can serve both the beginner who will need to approach things simply, the journeyman who is ready to move through several pieces at a time, and the expert who can complete the entire process without guidance.

2. Provide the user with control over how they progress through the components of the task. By letting the user set the pace and complexity for themselves, we provide just enough support for them regardless of where they are from a learning perspective.

3. Make support for the task obvious, relevant, and easy to find, but not required, intrusive, or generic. Think friendly help button with context-specific instructions not an interruptive, clueless Clippy. The user needs to know they are supported but not be required to use or even take time away from their task to dismiss the help. Make tutorials optional so that they can be accessed within the context of the task (but are not tutorials that launch automatically just because this is your first time logging on since it was created).

CHAPTER 7   THE EVOLUTION AND REVOLUTION OF PEOPLE

OH CLIPPY...

Clippit (more commonly referred to as Clippy) was an animated paperclip designed to be an Office Assistant for Microsoft Word in 1997. It would pop up at inopportune times while the user was working and chirp "It looks like you're trying to write a letter – would you like some help?" The overwhelming reaction from users was, unsurprisingly, abysmal. Termed the "one of the worst software design blunders in the annals of computing" by the Smithsonian,[8] it was removed from the product in 2007. Even focus groups exposed to Clippit before release disliked the chipper paper clip. (Did we mention about the importance of listening to your research results?)

Cultural Considerations

Anyone who has traveled outside of their hometown, especially abroad, has experienced the differences that a new location may offer. It can be as simple as whether the iced tea you order is sweet or whether it is iced at all. Or it can be as complex as how to greet someone you've just met. The differences between cultures affects not only our interactions with other individuals but how we, as individuals, interact with things.

In the early 1970s, an IBM scientist by the name of Geert Hofstede published a study based on an enormous survey of over 100,000 people from countries all around the world. The study outlined four cultural markers – comparison scales which described an aspect of culture and attitude.

---

[8] Conniff, Richard. "What's Behind a Smile?" Smithsonian Magazine, August 2007 pp. 51–52.

CHAPTER 7   THE EVOLUTION AND REVOLUTION OF PEOPLE

- **Individualism–Collectivism (rule-based vs. holistic)**: This dimension deals with the interdependence within a society and how people view themselves within (with an "I" or a "we" focus). A high score in this dimension indicates an individualistic society in which the individual is more of a focus than the society. A low score suggests a focus on society and one's place within that larger group.

- **Uncertainty Avoidance**: This dimension reflects how a society deals with the uncertainty of the future and the lengths to which they go to try to control the uncontrollable. High scores in this area reflect a greater fear of the unknown and generally suggest a higher level of rules, order, and rigidness to attempt to control it. Lower scores show a society more willing to accept the unknown and be comfortable with it.

- **Power Distance (Strength of Social Hierarchy)**: This dimension describes how comfortable a society is with a disparity in power. In a culture ranked high in power distance, inequalities are acceptable and viewed as the natural order of things. In a low power distance culture, all individuals are viewed as equal, and instances of inequality are viewed poorly.

- **Masculinity–Femininity (Later Changed to Motivation Toward Achievement and Success)**: This dimension describes what is valued more highly by a society with regard to achievement – individual accomplishments or societal betterment. High scores reflect a "decisive" or accomplishment-focused

CHAPTER 7   THE EVOLUTION AND REVOLUTION OF PEOPLE

society that views competition and being the best as the highest forms of achievement. Low scores reflect a "consensus" or empathy-focused society that views well-being or helping others as the admired forms of achievement.

Later, with additional research, more dimensions were added to provide a fuller picture:

- **Long-Term Orientation**: This dimension deals with a society's connection with its own past and willingness to adopt new methods and rituals. A high score indicates a very practical approach to traditions and willingness to change and update approaches as new learning is incorporated. A low score suggests a country that values traditions and views new developments with hesitation and fear.

- **Indulgence**: This dimension reflects the degree to which people have been raised to control their desires and wants. A high score in this area suggests a society willing to give in to temptation – to indulge themselves. This group values leisure time and views it as a right. A low score is associated with societies that look down on self-indulgence and instead admire restraint.

Hofstede's work has been influential in helping people from different cultures understand each other's motivations and approaches.

CHAPTER 7   THE EVOLUTION AND REVOLUTION OF PEOPLE

DO NOT GET THE MEAT POWDER

I used to travel to China regularly to work with my design team there. China has a relatively high power distance, which showed up in really interesting ways. When I was in the United States, it showed up as an unwillingness to contradict me (the boss) or point out flaws in my approach. However, when I was in China, it shifted once we left the office, and I was still "the boss" but also inexperienced in local customs and preferences. At one point, a group of us went to a street vendor selling what appeared to be a form of tacos. I asked about it and was told it was a Chengdu specialty (everything was a "Chengdu specialty" that was edible) called Ban Chang Kuih, a kind of treat that consisted of a small pancake with a filling of your choice such as jam, sweet cream, meat powder, and so on. My guide, one of our designers, suggested I get the sweet cream and meat powder as that was his favorite. I demurred and said I'd go with the sweet cream and jam because...well, meat powder is not the most appealing sounding thing, even though I'm typically very adventurous in a culinary setting. Another American designer with us also said he'd like the sweet cream and jam. Our guide agreed and got us the pancakes. My companion got sweet cream and jam and I got...sweet cream and meat powder. He told me "I know you'll like this better." This typified my interactions with the team whenever they were in an area of perceived greater knowledge. As you can see, the way cultural dimensions are displayed can be complicated and nuanced. And, for the record, do NOT get the meat powder.

But how does this affect design? Well, when Hofstede's work first came out, there was a push to try to customize experiences for different cultures. Amazon and other big consumer-facing tech companies experimented with providing different experiences for different countries based on the cultural dimensions. For example, a country ranking high on individualism–collectivism might provide a more personalized experience focused on the individual while one ranked low on that dimension might

172

CHAPTER 7   THE EVOLUTION AND REVOLUTION OF PEOPLE

focus on how the product is part of a large societal context. One enduring example of this is McDonald's. Check out the different ways McDonald's designs its websites for different countries:

Brazil:

China:

CHAPTER 7   THE EVOLUTION AND REVOLUTION OF PEOPLE

Sweden:

Germany:

CHAPTER 7   THE EVOLUTION AND REVOLUTION OF PEOPLE

The differences (and similarities) in the design choices are fascinating. And effective. While maintaining a region-specific website may be cost-prohibitive for many companies, for something as personal and consumer-responsive as food, it's a necessity.

The "We" in West Should Be Replaced with an "I"

Subsequent to Hofstede's early work, numerous studies over the past 50 years have highlighted how Eastern cultures – those predominantly in Asia – perceive and think about the world differently than those from the West. People from Eastern cultures tend to see the world more collectivistically, meaning that they process information based on more than just one individual or feature (sometimes called "holistic" processing). Contrarily, those from Western cultures (e.g., Europe and North America) focus more on singular elements that are shared (sometimes called "rule-based" processing) or on the salience of an individual rather than the group or context in which an individual may be embedded.

To illustrate what this looks like, we can reference a study where research participants were shown a flower (the target object) and asked which of two groups it belonged to. In Group 1, though all of the objects did not share any single feature, *in general*, they were more alike than different. All of the flowers in Group 2, however, had one feature (and only one feature) in common: a straight stem (which the target object also possessed), but other than that one feature, they didn't resemble one another holistically as much as those of Group 1.

When presented with these two groups, roughly two-thirds of European-Americans (i.e., those from Western cultures) indicated that the target object belonged to Group 2 (the rule-based/individualistic group), based on the configuration of its stem. Conversely, over 60% of individuals from East Asia (studying in the United States) and the majority of Asian

175

CHAPTER 7   THE EVOLUTION AND REVOLUTION OF PEOPLE

Americans (i.e., descendants of an Eastern culture who were born in the United States but raised by parents from the East) indicated that the target object belonged to Group 1 (the holistic or family resemblance group) because it had more features in common with all of the members of that group, taken as a whole, than with Group 2.[9]

These differences are not just artifacts of a laboratory setting, either. Comparisons of marketing strategies in Asia and the United States have provided ample support for the validity of these research findings. For example, a comparison of successful ad campaigns (for the same product) in the United States and Korea showed two very different approaches

---

[9] Nisbett, R. E., & Miyamoto, Y. (2005). The influence of culture: holistic versus analytic perception. *Trends in cognitive sciences*, 9(10), 467–473.

CHAPTER 7   THE EVOLUTION AND REVOLUTION OF PEOPLE

(as was illustrated previously with the various McDonald's ads). In the United States, the focus was on the benefit brought to the consumer as an individual, which was highlighted using phrases like "you will love." On the other hand, in Korea, the same detergent was marketed as being something not that the individual (i.e., "you") would love but rather "the family."[10]

UNDERSTANDING EACH OTHER

At one point, I had two groups of designers working for me: One group in China and one group in the United States. The differences in approach were immediately apparent not only in the approach to design work (the China designers tended to surface everything in the interface in a single page to provide optimum choices to the end user, whereas the US designers provided a walled-garden approach to guide users through a hierarchy of choices) but also in communication about design requests. Initially, it could take three or four iterations before partnered designers could align and move forward. Teaching about Hofstede's cultural dimensions provided a lot of "aha" moments for the team and let them communicate more effectively. If you're interested in learning more about cultural dimensions, check out the country comparison tool by Culture Factor (`https://www.theculturefactor.com/country-comparison-tool`). By using this tool, you can compare and contrast different countries on the Hofstede dimensions.

Other ways to influence purchasing in collectivistic cultures is through peer reviews and online ratings of products. In a laboratory examination of consumers in Hong Kong (collectivistic) and Australia (individualistic),

---

[10] Shavitt, S., & Barnes, A. J. (2020). Culture and the consumer journey. *Journal of retailing*, 96(1), 40–54.

177

CHAPTER 7   THE EVOLUTION AND REVOLUTION OF PEOPLE

the likelihood of purchasing books online was influenced markedly more by peer reviews (i.e., those who had purchased the book previously) for individuals in Hong Kong than Australia.[11] Based on other analyses within the study, there was strong evidence that these results speak to an increase in the trust felt toward the books, which was engendered through the reviews of others from the society.

Uncertainty Avoidance and Its Impact on Evaluating New Products

In a well-designed study, Lee, Gabarino, and Lerman (2007) examined the role that the uncertainty avoidance (UA) index of a particular country would have on purchasing decisions related to novel products (i.e., products for which there is also uncertainty about its use, value, reputable suppliers/manufacturers, etc.).[12] In order to determine the UA of a given market, they utilized the updated analyses of nation UA performed in 2006 (as an update to Hofstede's 1984 study).[13] They found that the UA was a strong, significant predictor of individuals' openness to considering a new product that would be introduced. Thus, when thinking about a new market or segment, though it is worthwhile to leverage existing, familiar, and trusted processes, imagery, and brands, these alone may not be

---

[11] Sia, C. L., Lim, K. H., Leung, K., Lee, M. K. O., Huang, W. W., & Benbasat, I. (2009). Web Strategies to Promote Internet Shopping: Is Cultural-Customization Needed? *MIS Quarterly, 33*(3), 491–512. https://doi.org/10.2307/20650306.

[12] Lee, J. A., Garbarino, E., & Lerman, D. (2007). How cultural differences in uncertainty avoidance affect product perceptions. *International Marketing Review, 24*(3), 330–349.

[13] Rothaermel, F. T., Kotha, S., & Steensma, H. K. (2006). International market entry by US internet firms: An empirical analysis of country risk, national culture, and market size. *Journal of Management, 32*(1), 56–82.

CHAPTER 7 THE EVOLUTION AND REVOLUTION OF PEOPLE

sufficient to entice users to a particular product. Indeed, in the case study that introduces their paper, Lee et al. talk about the long road required for EuroDisney (located in Paris, France) to reach not just viability but profitability.[14] Put differently, one should never assume that just because you build it, "they" will come. Regardless if the "it" is a shiny new website or a gleaming Le Château de la Belle au Bois Dormant (i.e., Sleeping Beauty's Castle).

The Yin and Yang of Purchasing Decisions

Cultural differences related to motivations can be thought of as those that relate to an achievement focus (which not coincidentally is also related to individualism) and what is thought of as a maintenance or loss prevention focus (which is also related to collectivism). When considering these two orientations, similar results (i.e., increased purchasing of services) can be obtained as long as the messaging is tailored to the orientation.[8] For example, framing the value of paying extra for expedited shipping was successful in achievement-focused cultures (i.e., the West) by prompting consumers with notes that appealed to a gain – in this case that they would be able to start enjoying their book sooner if they paid extra. However, for Eastern consumers, the most effective messaging wasn't about enjoying the book sooner but rather that without expedited shipping "they would have to wait longer," which represents a loss rather than a gain. These differences, though subtle, have been replicated enough times in both laboratory and real-world settings to warrant serious consideration when designing for global audiences.

---

[14] https://www.forbes.com/sites/carolinereid/2023/07/06/the-secrets-of-disneys-17-billion-theme-park-profits/

CHAPTER 7 THE EVOLUTION AND REVOLUTION OF PEOPLE

Importance of Personas (or Are They Personae?)

Before we debate the linguistic accuracy of personas vs. personae (don't worry – we won't), let's take a moment to sort out all the terms that get conflated when talking about this concept.

Persona: A persona is a *representation* of a user of a product created from data gathered through interviews and observations. Persona can help you by keeping a relatable abstraction (usually of the construction "Label + Name" like "Trendsetting Maria" or "Grammatically Correct Frank") of your user in front of the team when solving problems throughout the product development life cycle.

What a Persona Isn't:

- An actual individual.

- A Wikipedia/encyclopedia entry.

- Based on fabricated or anecdotal information – this last piece is the most commonly committed sin of persona development. Usually a group of people (or, shudder, one person) decides that they "know" who their users or customers are and puts together a list.

Segment: A segment is a subset of buyers with similar characteristics identified from data gathered through market analysis. Segments let companies target groups of customers with specific products and marketing techniques.

What a Segment Isn't:

- An actual individual

- An entire market

- Based on a news article in a business publication (note the "gathered through market analysis" from above)

180

CHAPTER 7 THE EVOLUTION AND REVOLUTION OF PEOPLE

Archetype: An archetype is a stereotypical character derived from one of twelve Jungian archetypes which represents your brand using a human character such as Creator or Every Man. Archetypes are used to create an immediate relationship between a company and the audience to communicate the value and mission of the company.

What an Archetype isn't:

- An actual individual

- A user type

- Based on a stereotype

Role: The job that the persona does at a given time.

What a Role Isn't:

- An actual individual

- A job title

- Based on a permission group

Creating and Using Personas: The Right Way

Remember that a persona is an *abstraction* of an individual. Part of what makes personas useful is that they aggregate various problems, settings, attitudes, and roles that a large number of users experience into a single "user" (but again, not an individual) for whom you can begin to think through solutions.

If you have some understanding of the use cases and potential roadblocks your users may have, possibly due to experience or feedback that you've received, then you can begin to create lists of questions that you'll want to collect more data on so that you can understand if they are universal issues or those relegated to only specific subsets of your user base. For example, it may be that you know that principals use your

181

CHAPTER 7 THE EVOLUTION AND REVOLUTION OF PEOPLE

system differently than teachers (or bulk purchasers differently than retail consumers, etc.), so you would try to understand the similarities and differences between those individuals and also determine whether there are subsets within each category (e.g., school administrators and educators).

Now, if we didn't know where to start or what qualities might differentiate one group of users from another, then we would use an *inductive method* to *identify* the themes and patterns that might exist for a given group of people. Inductive reasoning takes place through qualitative methodologies, such as interviews or focus groups. Inductive methods provide one method to establish an hypothesis, but – IMPORTANTLY – not the means to test or validate it. This means that a focus group or interview should never serve as the only means for establishing user personas. This is because

- Qualitative methods do not analyze large enough sample sizes to be representative.

- Without proper training and experience, biases can be inadvertently introduced through the analysis and interpretation of qualitative data by novices.

- And, most importantly (and with an explanation that exceeds the scope of this text), the act of induction (unlike deduction) can have a false conclusion that follows from true premises.[15] For example, maybe the first candy you've ever tasted is coated extensively with malic acid, making it extremely sour rather than sweet. If you then took this specific instance and applied it generally, you would incorrectly presume that *all* candy is sour.

---

[15] Harman, G., & Kulkarni, S. R. (2006). The problem of induction. *Philosophy and Phenomenological Research, 72*(3), 559–575.

CHAPTER 7   THE EVOLUTION AND REVOLUTION OF PEOPLE

Indeed, given the nature of inductive methods, you technically can't say anything about "all," "most," or anyone outside of the individuals whom you've observed with any certitude. This is why inductive methods are not generalizable. However, they can provide a useful means for generating hypotheses to be tested. Just because those data don't allow you to make generalizations doesn't mean that the trends you see aren't generalizable but rather that you can't know if they are or aren't until you test using deductive methods.

Thus, once we have identified the user characteristics that we *think* matter (whether from past experiences or through inductive methods – see above), we need to use a methodology that will actually allow us to determine if this hunch/opinion/hypothesis holds up in the real world. This approach then relies on deductive reasoning – that is determining if general principles (i.e., your hypothesis) apply to specific circumstances (i.e., actual human users). Deductive reasoning uses quantitative measurement and statistical methods to determine the likelihood that the general principle actually applies to the specific circumstance being examined. Thus, any time we want to know if a change, like a refreshed home page, yields more of something (like clicks or conversions) or fewer/less (like error messages or time to navigate), we are conducting quantitative testing to make that determination. And the same holds true when trying to determine whether our users fit into various categories/ personas.

So how do we generate quantitative data about our users? The best way to do this is to gather information about those characteristics, behaviors, and qualities of our users that we believe will provide us with the right information to understand the specific problems they might encounter (which our designs will solve). This is where proper survey design can yield robust data that will allow you to (1) test whether your hypothesis/es were true and (2) see whether any of the characteristics, problems, tasks, etc. that you've identified as general areas of interest *cluster* together in interesting and useful ways.

CHAPTER 7   THE EVOLUTION AND REVOLUTION OF PEOPLE

"There aren't enough hours in the day."

Name: Vera Founder

Job Title: Owner/CEO

Age: 43

Preferred Device: iPhone (iOS)

About: Vera is pursuing her dreams of owning her own business and drinking only the finest beers by starting her own micro-brewery. She is obsessed with the details about her business and is constantly looking for ways to streamline and save on both time and money.

Context: She is required to spend much of her day traveling to local businesses to establish and maintain relationships for the distribution of her micro-brews.

Motivator(s): She not only wants to have a successful business but wants to "be there" for her employees.

Job Tasks:
- Establish and maintain sales pipelines
- Coordinate deliveries, purchasing, and distribution
- Manage on site employees
- Coordinate with brewmaster and general manager on scheduled events
- Troubleshoot any and all issues

Friction/Pain Points:
- Has to use multiple platforms to manage employee schedules, delivery and distribution times, and event calendars
- No "early alert" system for dwindling stock

Preferred Apps: WhatsApp, Instagram

Now that you have a persona, what are they good for? Personas can help you

- Envision and empathize with the person who is using your software (What makes them happy? Angry? What do they NOT care about?)

- Test ideas (Does this concept fit with their technological proficiency? Does it support their primary tasks? Does it solve a problem for them? Is it a big problem?)

- Keep the user represented throughout the product life cycle

- Inform and focus your journey map (remember our dog walking journey maps from Chapter 6?)

- Write user stories in an agile product development process

- Define problems (spoiler alert! we'll talk more about this in Chapter 12)

CHAPTER 7   THE EVOLUTION AND REVOLUTION OF PEOPLE

Personas help you ensure you are solving the right problem for the right person in the right way, which is why it is critical that they be based on good research. Because they are used in so many parts of the process, making up your personas based on a couple of customers, hearsay, or a blog article can have a snowball effect – doubling down on mistaken assumptions until they lead the product completely astray.

Recap

- While age is often considered an important attitude/cultural differentiator by the media, generational attitudinal differences are largely anecdotal and not particularly useful. Age-related physical differences are a more important design consideration.

- Differences in experience and expertise can make creating a one-size-fits-all product difficult. Use scaffolding to help beginner users without interfering with experts.

- Cultural considerations are important in communication, design, and product marketing. Be aware of the cultural dimensions in your target audience to better understand motivators and avoid missteps.

- Personas are research-based abstractions of the end user used to inform all stages of the product development life cycle.

Before You Go...

When Dr. Baker's kids were younger, she was struck by how they interacted with things very differently than she did at that age. Any flat screen was immediately covered in fingerprints because the assumption was that all

CHAPTER 7   THE EVOLUTION AND REVOLUTION OF PEOPLE

screens were touch screens. And it wasn't just her kids – a friend of hers told a story of his twins putting on a play for him and his spouse. The first thing the boys started doing is spinning a glow stick on a string in a circle. When asked what they were doing, they said "this is the show loading for you!" Seeing a spinning loading icon or touching a flat surface to make a selection was natural to them because it was an integral part of their day-to-day experiences. With observations like this, it's easy to see how the popular press has promoted the idea of "digital natives." But is that entirely accurate?

The straightforward answer is "no."[16] It turns out that when you dig just beneath the surface of what seems like commonsense (i.e., kids today have been around screens their entire lives, so they must be better at them), the studies just don't support it.[17] Indeed, there is a difference between mimicking how something works and understanding it. For example, small children also know that "choo choo" is the sound that trains make, but that doesn't mean that they're qualified to be a conductor or lecture on the physics responsible for the advent of the steam locomotive. What's more is we are now starting to see evidence that the idea that children are "digital natives" can even have counterintuitive impacts on their learning, whereby too much technology or its introduction at young ages can actually impair rather than enhance the learning of the children who continue to be born into an increasingly digital world.[18]

---

[16] Prensky, M. 2001a. Digital natives, digital immigrants. On the Horizon 9 (5): 1-6. http://www.scribd.com/doc/9799/Prensky-Digital-Natives-Digital-Immigrants-Part1.

[17] Helsper, E. J., & Eynon, R. (2010). Digital natives: where is the evidence?. *British educational research journal*, *36*(3), 503–520.

[18] OECD/Rebecca Eynon (2020), "The myth of the digital native: Why it persists and the harm it inflicts", in Burns, T. and F. Gottschalk (eds.), Education in the Digital Age: Healthy and Happy Children, OECD Publishing, Paris, https://doi.org/10.1787/2dac420b-en.

CHAPTER 8

Communication Is Hard (and We Suck at It)

JLR: I know some linguists hate it, but I still get a chuckle every time I see "GHOTI" referenced.

RB: Ah, yes, and by "GHOTI," you mean a nonsense word that is pronounced "fish"?

JLR: Yep. That's the one. Of course, the joke is that to get this, we need to use the sound that "GH" makes in the word "rouGH," the "O" sound from "wOmen," and the "TI" sound from "palaTIal," which is technically correct in this Frankenstein's-monster-creation.

RB: Yes, but as offended linguists and English teachers would tell you, those sounds are taken out of context...

JLR: Yes, yes, but the point is the same: namely, that there are so many factors that we must take into account – subconsciously – when we read and speak English and that all of these are compounded by the fact that English is one of the least consistent languages on the planet.

RB: Am I to assume that you have another example (I know you do).

JLR: Of course I do, but this one requires you to say aloud the following four words: "LAMP," "RAMP," "VAMP," and "STAMP."

RB: Ok, I'll play along and do that. They all rhyme, so what?

© Jerome L. Rekart and Rebecca Baker 2025
J. L. Rekart and R. Baker, *Designing for Human Intelligence in an Artificial Intelligence World*,
https://doi.org/10.1007/979-8-8688-1418-1_8

CHAPTER 8   COMMUNICATION IS HARD (AND WE SUCK AT IT)

JLR: Well, given all of that, shouldn't the word "SWAMP" be pronounced differently?

RB: I shouldn't have indulged you.

JLR: [spells out the word letter by letter] Gnaugh, I would have shared regardless.

RB: What?

JLR: Gnaugh...pronounced using the sound that "GN" makes in "gNome" and that "AUGH" makes in "nAUGHt."

RB: Ugh. As in "UGH."

JLR: Fair enough.

Language: From Signals to Information

It is actually quite amazing that we are able to accumulate, understand, and use the complex systems that underlie language. When we are fluent in said languages, we can even do so with little expenditure of cognitive resources (think about the signs you randomly and without intention read while driving). Indeed, we even have the capacity to develop new ones that relate to one another and follow their own sets of rules, like HTML, Python, R, or SQL. This proclivity for language begins at an early age, when most humans are primed to take incoming sensory information, such as sounds made by those around them, and look for what have been called "meaningful signals"[1] within them. For example, a young child repeatedly hears the sound of their own name and begins to associate that particular sound with themselves. Even more abstract ideas like "more" are assimilated within the developing lexicon of a child. These ideas can

---

[1] Pinker, S., & Jackendoff, R. (2009). The components of language: What's specific to language, and what's specific to humans. *Language universals*, 126–151.

CHAPTER 8   COMMUNICATION IS HARD (AND WE SUCK AT IT)

then be applied to novel circumstances. Thus, although young Timmy always heard "more" used in the context of "juice" (which resulted in its reappearance), he soon notes that it also seems to bring additional fruit. When he is finally able to utter the sound himself, he tries it out after finishing a delicious, flaky treat (the tall one called it a /kook-ee/) and wow! That worked. Now that he has found that he was correct and that his usage of that sound results in the appearance of another /kook-ee/ he begins to look for other transferable concepts.

SIGNAL VS. DATA VS. INFORMATION

Let me geek out a moment on the concept of the difference between and the relationship of data, signals, and information. Data are the raw material – the ones and zeros if you're a computer geek, the words of a story if you're a language geek, the numbers if you're math geek. Data in and of itself is meaningless – it is simply how we represent discrete bundles of stuff that we use to communicate. Signals are the medium we use to transmit data from a source to a receiver – your voice, the Internet, the ink on the page, the touch of a hand on your shoulder. Information is the meaning derived from combining the data received through a signal in a given framework or context – the volume of a sphere, the story, the bank account. So why do we care about breaking down these definitions? Understanding communication boils down to understanding how a very solid repeatable thing (data) loses integrity when sent through a less than perfect signal and interpreted (information). By breaking these things down, we can better understand (philosophically and mathematically) how to optimize our communications. If you're interested in more geekiness around this idea, check out Claude Shannon's information theory.

189

CHAPTER 8   COMMUNICATION IS HARD (AND WE SUCK AT IT)

As we age, we then begin to learn that we can also have visual symbols (i.e., letters) that stand for the auditory symbols (i.e., phonics),[2] which in turn stand for the things that we are trying to communicate. Thus, after a few years, slightly less young Timmy now knows that "cookie" (if he lives in the United States) is the visual representation of the sound /kook-ee/ that delivers the oh-so-sweet, flaky delectable treat. After several more years (and countless more cookies), Timmy will gradually gain fluency in his native language and by around the 6th grade will have achieved a working knowledge of at least 10,000 words.[3] Note that the "at least" used in the last sentence is a hedge because many language researchers cannot agree on what constitutes a word (e.g., are "cookie" and "cookies" two words or one word with two forms?), and an in-depth debate about this is not something that we need to engage in here (or really, anywhere).

This level of working fluency is accompanied by a complicated internalization of the rules of language. For example, Tim's brain (he doesn't go by "Timmy" anymore as he thinks it is a bit immature) can now finish sentences others are speaking before the speaker does based on what he has experienced previously. Indeed, though Timmy...sorry, Tim...may not think of himself as being mathematically inclined, his brain routinely performs a complex set of statistical inferences to identify likely words that will complete thoughts and concepts being conveyed by others.[4]

---

[2] We=individuals growing up in cultures with phonetic languages.

[3] Nation, I. (2006). How large a vocabulary is needed for reading and listening?. *Canadian modern language review, 63*(1), 59–82.

[4] Martin, C. D., Branzi, F. M., & Bar, M. (2018). Prediction is Production: The missing link between language production and comprehension. *Scientific reports, 8*(1), 1079.

190

CHAPTER 8   COMMUNICATION IS HARD (AND WE SUCK AT IT)

Along these lines you may have encountered (on the Internet, where else?) a passage that has letters in the middle of words scrambled, yet you can read it with little difficulty.

For example:

> "you could ramdinose all the letetrs, keipeng the
> first two and last two the same, and reibadailty
> would hadrly be aftcfeed."[5]

Although you can read this passage (as could Tim), to say that it could be done without readability being affected (or aftcfeed) is at best a bit of an oversimplification. This finding, which was based on a Ph.D. thesis published in the 1970s and subsequently reprinted due to the popularity of the meme,[6] relies on some tricks for it to work. To be sure, there is something neat about the fact that our brains are taking in not just the specific letter orders but that they are making broad statistical inferences that allow us to read. However, the fact that small words, like "the," "two," and "last" aren't scrambled also facilitates readability and provides some contextual "anchors" for our minds to latch onto when interpreting the rest. Thus, you likely would find the following passage to be much more difficult to read than the first – though again due to some rather complex statistical reasoning – still mostly readable:

> "rdianeg tsih psgasae in'st as slmpie bsaucee tehre are
> feewr ahorcns taht our bniars are albe to iteerpnt rldpiay."[7]

---

[5] Rawlinson, G. (1999, May 29). Letter: Reibadailty [Letter to the editor]. *New Scientist*, 2188.

[6] Rawlinson, G. (2007). The significance of letter position in word recognition. *IEEE Aerospace and Electronic Systems Magazine, 22*(1), 26–27.

[7] "Reading this passage isn't as simple because there are fewer anchors that our brains are able to interpret rapidly."

CHAPTER 8   COMMUNICATION IS HARD (AND WE SUCK AT IT)

It is through this same mechanism (implicit, unconscious application of statistical reasoning to language) that as he reaches adulthood, Tim's mind will process even more complicated information from statements and will be able to apply it in a way that allows him to identify underlying meaning and understand sarcasm, puns, and double entendres.

Production vs. Comprehension

As anyone who has ever tried to learn another language can attest, the ability to understand what other people say comes well before one feels comfortable or efficacious actually saying something back to them. In fact, this dissociation has some interesting repercussions that can (and should) be exploited in design. Though the underlying neural networks are in actuality much more complex than described here, a primary contributor to this dichotomy is that there is actually a separation of the hardware responsible for language comprehension (e.g., understanding; Wernicke's area) and language production (e.g., talking; Broca's area).[8] This is why you will often see or hear of stroke victims who have to relearn how to talk yet can understand whatever is spoken to them just as they could before the brain injury (and, though rarer, the opposite case can also occur).[9] It should be noted that this division also extends to other acts of producing language, as occurs between reading and writing.

---

[8] Foundas, A. L. (2001). The anatomical basis of language. *Topics in Language Disorders, 21*(3), 1–19.

[9] Sinanović, O., Mrkonjić, Z., Zukić, S., Vidović, M., & Imamović, K. (2011). Post-stroke language disorders. *Acta Clin Croat, 50*(1), 79–94.

CHAPTER 8   COMMUNICATION IS HARD (AND WE SUCK AT IT)

WHY WON'T IT UNDERSTAND ME?

However, this knowledge doesn't make it any less frustrating! As much as I'd like to assign blame for my lackluster Mandarin pronunciation to a poor user interface (it isn't), shoddy auditory recognition software (it isn't), or any number of other reasons (they wouldn't be valid), the truth is that when I attempt to speak Mandarin to that green little owl in my language app, I am doing so several levels lower/more remedial than when I interpret what the system is saying to me.

Indeed, numerous imaging studies of the brain in action have shown that different brain regions are recruited to process various modes of communication, for example, as is seen with the division of labor underlying the processing of sentences alone compared with those from a conversation or discourse.[10] Consequently, we cannot think of communicative language as a singular, monolithic process but rather as two interrelated processes that both depend on one another and, in mature individuals, can be independent. If that last sentence seems convoluted, it is because the system it is trying to describe is exactly that: a complex, hard-to-simplify form of neurological input and output that we all take for granted.

Another byproduct of this neurological wiring – and one that should be remembered in design contexts – is that the brain has an actual preference for putting easy things first. This "easy first" facilitation has been demonstrated in the following ways[11]:

---

[10] Gernsbacher, M. A., & Kaschak, M. P. (2003). Neuroimaging studies of language production and comprehension. *Annual review of psychology, 54*(1), 91–114.

[11] MacDonald, M. C. (2013). How language production shapes language form and comprehension. *Frontiers in psychology, 4*, 226.

CHAPTER 8 COMMUNICATION IS HARD (AND WE SUCK AT IT)

- When forming sentences – either written or spoken – people place less complex/more frequently used words at the start because they are easier to pull from long-term memory.

- More "ums," "ers," and other linguistic placeholders occur later in sentences than earlier.

- When ordering events, directions, or items, people will place the ones that are easiest to articulate, reach, and or obtain first.

The last finding, though not directly related to language, is perhaps the best example of designing to make life easy for your users – not to impress them with

- Your perspicacity/size of your lexicon

- How you can make site navigation feel like an escape room ("Only 20 minutes left")

- Using jargon/slang/language that is specific to your site ("Click the flurg to gain a 10% off coupon")

Trigger Warning: Words As Emotional Cues

It seems as though people vacillate between two extreme positions when it comes to the emotionality of words. On one end of the spectrum is the old schoolyard chant that "sticks and stones may break my bones, but words will never hurt me," which suggests that we can be impervious to linguistic assaults. On the other are statements like "the pen is mightier than the

CHAPTER 8   COMMUNICATION IS HARD (AND WE SUCK AT IT)

sword,"[12] which is often used to suggest that great change, like political revolutions, may be ushered in through prose. After all, how do great works of literature and theater create tension and conflict but through discourse, debate, and discussion. Just think of the many high school students across the United States who will spend (or have spent) many class periods dissecting the pain and shame that can be evoked not with sentences or words, but with a single, scarlet letter.

So, do words carry any emotional weight, and if so, can we exploit that relationship to tap into people's psychological hardware?

Let's get it out of the way that, yes, words evoke emotions. As suggested above, we don't need to delve into psychological research to know this is so but rather only need to reference the entirety of the Western literary canon.[13]

Still, given the nature of this book and our backgrounds, we can't help but place the psychology of this effect in context. Over the past quarter century, researchers have created large lists showing that certain words *reliably* evoke the same feelings across people.[14] This should come as no surprise as some words like "scar" really don't have positive connotations and so would be expected to make people feel apprehensive and other words like "hero" have strictly positive connotations (either as someone who is valiant or the sandwich). Interestingly, regardless if the words are positive ("hero") or negative ("scar"), as long as they have some emotionality associated with them (relative to a neutral word, like "dozen"

---

[12] https://www.bbc.com/news/magazine-30729480

[13] You know, all that stuff that your middle and high school (and beyond) teachers tried to get you to read and think about?

[14] Bradley, M.M., Lang, P.J.: Affective norms for English words (ANEW): Instruction manual and affective ratings. Technical Report C-1, The Center for Research in Psychophysiology, University of Florida (1999).

CHAPTER 8   COMMUNICATION IS HARD (AND WE SUCK AT IT)

or "segment"), they have been found to unconsciously "grab attention," which results in them being processed faster (i.e., they get read faster) and they are remembered better.[15]

This means that words in headlines, ledes, and even category or section headings can affect how users feel when using a product. We have opted to use the generic term "affect" here and hesitate to use the word "impact" as that suggests something profound, and what you are most likely to do with proper word selection is nudge how your users are feeling. Though small in scale, several nudges in the right direction could successfully prime your users to be in a more open emotional state rather than one that might be more prone to boredom, frustration, or rage.

Note that we are talking about normal emotional responses here. This is different from what was originally intended by including so-called "trigger warnings" with texts or programs. Trigger warnings were originally implemented as a means of helping victims of trauma (and who suffer from posttraumatic stress disorder) avoid scenarios that could involuntarily "trigger" a flashback or strong feelings associated with the precipitating traumatic event. These warnings were never meant to be used to merely signal that something unpleasant might be stated (and indeed, such generalized [mis]uses may even have counterintuitive results, perhaps increasing the anxiety of individuals rather than assuaging it[16]).

---

[15] Kousta, S. T., Vinson, D. P., & Vigliocco, G. (2009). Emotion words, regardless of polarity, have a processing advantage over neutral words. *Cognition, 112*(3), 473–481.

[16] Bellet, B. W., Jones, P. J., & McNally, R. J. (2018). Trigger warning: Empirical evidence ahead. *Journal of Behavior Therapy and Experimental Psychiatry, 61*, 134–141.

CHAPTER 8   COMMUNICATION IS HARD (AND WE SUCK AT IT)

What You Heard Is Not What I Meant

Even when we share a language with someone, we can still manage to misunderstand each other on a regular basis, as anyone who has ever been married will attest. Communication is hard! But why should that be? After all, a rose by any other name would still be a rose. The answer lies in the complexities of perception, context, and interpretation. In his classic work, *Women, Fire, and Dangerous Things*, George Lakoff shows that how we categorize things reveals how we think and correspondingly how our perception of our environment in turn affects how we use and interpret language (symbols). In short, language taken at face value will conflate things (for example, the aboriginal language of Dyirbal has categories that include women, fire, and scorpions), and to meaningfully interpret it, you must include context.

This idea is especially important in the business world, whether communicating with the C-suite in the boardroom, with your cross-functional team in a standup, or with your users through your design work. If you do not have the same context – the same perception of what is intended to be conveyed – you will have miscommunications. This context could be physical or emotional, but without a common context, your message has a high chance of failure. For example, consider a learning module that, as part of teaching someone how to write instructions, asks you to explain how to operate a coffee maker. This might seem fairly straightforward to those of us who are coffee addicts in the United States. However, this particular learning module was sent to students in refugee camps, who had never seen a coffee maker, let alone needed to know how to operate one. While the language of the module was perfectly clear, the lack of context ensured that the students wouldn't learn what they needed.

CHAPTER 8   COMMUNICATION IS HARD (AND WE SUCK AT IT)

Design Systems As Language

You'll hear a lot of businesses talk about the importance of a design system. "Have you ever built a design system?" is a common question during design interviews. But what is a design system? Simply put, a design system is language – a set of grammar rules and vocabulary to communicate a design. The visual design (colors, icons, spacing, fonts, etc.), the interaction patterns (button behaviors, logins, error handling, etc.), and the components (coded units of the interactions units, the CSS, etc.) come together to provide this language that can be used by developers and designers to communicate a cohesive experience.

The idea of design as a language is not new. *A Pattern Language* by Alexander, Ishikawa, and Silverstein defined a set of architectural components and rules for combining them to be used for designing everything from houses to cities. They suggested not only the elements but how things would flow from a public to semipublic to private area, providing guidelines for transitions and movement. The idea of codifying these things was met with mixed reviews – some embraced the idea, while others felt it was too restrictive to the art of architecture. The larger a project and the more independent parts that needed to come together, the more useful the pattern language becomes. Fast forward to today, we find ourselves having the same arguments in software design. Designers sometimes balk at having to use design systems, protesting that it hinders their creativity. But much like architecture, the larger the system, the more important it is that you have a cohesive set of rules to ensure the experience of using them is seamless.

It might be tempting to see a design language as an easy fix: define the fonts, colors, and icons, create some components as you go, and anyone can apply that to every product in your portfolio – no matter when or how they were created. This was, in fact, part of the intent of the original pattern language – to make architectural elements accessible to the layman. However, just like any language, knowing the grammar and vocabulary

CHAPTER 8   COMMUNICATION IS HARD (AND WE SUCK AT IT)

is not the same thing as being able to speak fluently. And translating something from one language to another often results in nonsense due to the lack of context. Amazon experienced this in 2020 when launching its Swedish website. Faulty translations included mistakenly using the Swedish word for male genitalia instead of the word rooster[17] and the Nintendo Switch became the Nintendo Circuit Breaker.[18] While hilarious, the results were predictably disastrous for sales.

Design languages are useful as a consistent, complex, ever-evolving record of how systems should communicate with the user and how new designs and components can flow with to and from the system. They are fantastically useful for large-scale systems where it can be difficult to be consistent and the economy of scale makes the efficiency of component reuse attractive. But they are not an easy fix for consistency issues, and they cannot be used simply as a skin to make disparate systems work together that were not designed from the foundation up to do so. Design systems require significant work to set up and maintain.

Effective Use of Design Systems

So, how DO you use a design system effectively? Well, like any good engineer will tell you when you ask a question on how to do something – it's complicated. By that, we mean it requires an adaptable understanding of the problem to be solved (see Chapter 12 for more on that) and the context in which it is being used which will be unique to your situation.

---

[17] https://www.reuters.com/article/business/lost-in-translation-amazon-website-launch-trips-over-faulty-swedish-idUSKBN27D2EB/

[18] https://www.dailymail.co.uk/news/article-8896521/Amazons-Swedish-website-launch-laughing-stock-poor-translation-algorithm.html

CHAPTER 8   COMMUNICATION IS HARD (AND WE SUCK AT IT)

However, there are some tips we can offer on how to effectively use these systems, understanding that you need to be flexible when creating and using them:

- **Consider Both Design and Development.** A design system that only considers the aesthetics and/or behavior of an element but not the interaction with other coded elements will never get used (or will be repeatedly "customized" – which is code for "rewritten"). Similarly, elements that only consider the code interactions and are optimized for ease of coding will ignore how elements move or interact with one another from a user's perspective which will, once again, require "customization." Once an element is rewritten, you lose the efficiency of reuse as well as the ability to update the element when new code is available.

- **Think Scale.** A great benefit of design systems is the ability to scale your system. Updates to elements can be done *en masse*; initial coding is done once for each element then reused, and quality assurance has a consistent way to test adherence to standards. As such, when creating elements, they need to be as broadly applicable as possible. Consider different scenarios in which the element could be used and plan accordingly. Consider the implications of having multiple instances of the same element on a single page. Consider how it might be used across different product lines. And consider all of these things, not just during creation, but when updating elements.

200

CHAPTER 8   COMMUNICATION IS HARD (AND WE SUCK AT IT)

- **Enforce Adoption.** It may seem that this goes without saying, but if you are going to invest in standing up a design system, mandate its use. You will reap none of the benefits of the system if it does not get used, and it is human nature to resist change. And resist they will! Adopting a design system will inevitably be shuffled to the bottom of any list as new features and bug fixes take priority for short-term benefits. Letting this happen will prevent you from seeing the benefits of your (significant) investment and will ultimately lead to the slow and painful death of the design system. Consider something similar to the Bezos API mandate.[19] As they grew and acquired new products, Amazon was faced with a growing set of disparate systems that were unable to communicate with one another in spite of repeated requests for the systems to prioritize open communication. Finally, in 2002, Jeff Bezos famously issued a mandate that all products would expose their functionality through service interfaces and would only talk to each other through those interfaces (rather than through hard-coded links) and that anyone who did not do it would be fired. While harsh, the mandate was effective, and now, Amazon is able to scale and grow effectively and efficiently.

---

[19] https://nordicapis.com/the-bezos-api-mandate-amazons-manifesto-for-externalization/

CHAPTER 8   COMMUNICATION IS HARD (AND WE SUCK AT IT)

Storytelling

A story is a story, whether presented between two covers, or on a screen. If the words have dramatic impact, if the pictures are visually appealing, if the theme is emotionally relevant, then certainly it is worthy of a reader's attention.

—Stan Lee, December 1968

Let's take a moment to define the concept of a story. If I were to ask a robot to describe a story, they would likely tell me it is a recounting of a series of events, either real or imagined. But if I ask a human, they will tell you something different – and that something will be personal, colorful, and interesting. Stories are much more than a simple recording of the video conference meeting you attended last Tuesday. They provide the context of how people came to the meeting, gloss over the uninteresting details of he said/she said, and highlight (and embellish) the drama of when it was revealed that the funding for the project was approved. Stories are an interpretation of events that are heavily overlaid with context and understanding. And as such, stories are not just useful for entertainment – they can provide an excellent way to communicate information, ideas, and events in a way that the audience can understand the perspective of the teller.

Stories let us see things from the storyteller's perspective and compare how that perspective is different (or similar) from our own. Stories are a particularly powerful tool when working in design – you can use stories to help stakeholders understand the pain points you are trying to address, to help your development staff clarify the problem they are trying to solve, to help your marketing department create a campaign that connects with users, or to help your support department answer calls with a deep understanding of the client's situation.

CHAPTER 8   COMMUNICATION IS HARD (AND WE SUCK AT IT)

AND THE PRIZE FOR BEST TYPE OF READING GOES TO...

Often, we think of stories as being either oral or written narratives – but they are more than that. Stories can consist of anything used to communicate: words, pictures, or sounds. A story could be a TikTok video, a manga, a podcast, a site visit report, a rock song, or a ballet. No matter the medium used for the story, they all provide an interpretation of a situation including the storyteller's perspective and reaction to it. I know a number of fellow parents that bemoan their child only reading manga or listening to audiobooks. But when it comes down to it – does it matter? The short answer is – no, not really. A recent study mapping areas of the brain activated by listening vs. reading indicates that the areas stimulated are highly correlated.[20] Put another way – reading or listening makes the same parts of your brain light up. So, go ahead and listen to your favorite romantasy while driving in rush hour, and know that it's just as effective. What about those comic books (sorry – graphic novels)? Turns out that they also engage our brains in complex ways.[21] It simply depends on what appeals to you personally.

As a communication tool, stories are incredibly powerful. Storytelling can enhance engagement, recall, and cross-cultural understanding. Stories provide context and direction for users that help them relate to the tasks and content more readily and remember sequences more easily. Incorporating narratives and storylines in your workflow makes tasks more engaging and to give users a sense of purpose and context. There are many reasons to use stories in your daily work.

---

[20] https://www.jneurosci.org/content/39/39/7722
[21] https://www.sciencedirect.com/science/article/abs/pii/S0079742119300027

CHAPTER 8   COMMUNICATION IS HARD (AND WE SUCK AT IT)

Improving Understanding

Stories can be particularly useful in helping bridge cultural gaps within design groups. When working on reservation software for daycare in recreation centers, Dr. Baker's design team in China struggled to understand the how/why/when for parents trying to make reservations for their children – leaving your children with a nonfamily member wasn't something they were familiar with. By creating stories about a typical experience describing everything from the tearful drop-off of a new parent to the frazzled attendance-taking of the overwrought teacher to the frenzied pickup of the parent, she was able to work with her team to picture the real pain points (being able to provide security and reliability in a distraction-laden environment) and address those, instead of focusing on less important design elements such as adding aesthetic improvements or enhanced reporting. What is particularly cool is to think that studies have even shown that narratives can create a convergence of neural activity across disparate people when they interact with, relate, and remember stories they've heard: thus bringing them together at a neurological level.[22]

For example, researchers viewed similar levels and locations of brain activation (in what is known as the DMN, or "default mode network") in English speakers who heard a story in English as did Russian speakers who heard the same story in their mother (Russia) tongue – thus, the commonality was driven by the actual meaning of the story. Similar results were seen between individuals who read a screenplay and those who actually viewed the scene that was depicted in the script![23]

---

[22] Suzuki, W. A., Feliú-Mójer, M. I., Hasson, U., Yehuda, R., & Zarate, J. M. (2018). Dialogues: The science and power of storytelling. *Journal of Neuroscience, 38*(44), 9468–9470.

[23] Tikka, P., Kauttonen, J., & Hlushchuk, Y. (2018). Narrative comprehension beyond language: Common brain networks activated by a movie and its script. *PLoS One, 13*(7), e0200134.

204

Enhancing Engagement

Stories enhance engagement by helping personalize the user's experience – making it their own. From a gamification perspective, you can leverage this through narrative-driven challenges (quests) that provide a storyline or context for users, making tasks more engaging and meaningful. For mundane, but practical, approaches, you can use narratives in instructions on how to use a product, in the marketing/sales enablement literature to provide case studies, and as graphics or sounds within the interface. Remember, stories don't have to be long narratives! A single picture can tell a story that engages the users when and where you need them, for example, an animated icon of a burning candle dripping wax for the passage of time or a person climbing a mountain to represent goals.

Improving Storage

Our brains are wired to hold on to stories. When we experience a story, multiple parts of our brain are activated and engaged. Our brain simulates what it's like to experience what is being described in the story, engaging

CHAPTER 8   COMMUNICATION IS HARD (AND WE SUCK AT IT)

our emotions and visualizations.[24] And for bonus points, stories provide our brains an immediate structure (at least the good stories do – this does not apply to the long rambling narrative your uncle recited to you at dinner describing his trek to the local fishing store to purchase a new lure) that can be used for storage and organization. These anchors then provide information that facilitates later retrieval and recall (as will be discussed in our discussion of cognitive cartography in Chapter 9). That information is now personal, engaging, organized, and, well, memorable.

Recap

- Language is a constructive process that relies on our brain constantly hedging bets about what will come next out of our or others' mouths.

- There is a temporal and a physical disconnect (i.e., different brain regions) that occurs when we are trying to understand what someone says and when we are actually saying something to them.

- Some words have psychological impact on our emotions and can be used to prime our brains for certain actions.

- Context is important in communication – and communication is hard…

---

[24] Lee, H., Bellana, B., & Chen, J. (2020). What can narratives tell us about the neural bases of human memory?. *Current Opinion in Behavioral Sciences, 32*, 111–119.

CHAPTER 8   COMMUNICATION IS HARD (AND WE SUCK AT IT)

- Design systems are a language that can be used to communicate effectively within a company and with users of a product.

- Design systems are useful when creating products at scale but are not a cheap panacea for inconsistencies and poor design. They require investment, commitment, and ongoing maintenance to reap benefits.

- Stories are a way for people to understand our perspectives more easily and can be used effectively in enhancing engagement, memory, and communication.

Before You Go...

Dr. Baker loves learning about holiday traditions around the world – the variety of celebrations is fascinating! Her favorite by far is one she has adopted in her home from Iceland: Jólabókaflóð (Yule Book Flood). On December 24th, families in Iceland exchange books and spend the rest of the evening reading their new books, sharing with each other what they are reading, and drinking chocolate. As a confirmed bookwyrm and chocolate fan, she would be hard-pressed to conceive of a better holiday tradition. Dating back to 1944, the tradition sprang up after Iceland gained independence from Denmark. In the postwar era, paper was one of the few things not rationed, so books became a popular gift. The tradition has grown since, with a yearly book bulletin (*Bókatíðindi*) issued to every household that people can use to order books for their friends and loved ones.

CHAPTER 9

I Remember When... or Do I?

JLR: Did you know that as people age they experience a positivity bias?

RB: That would seem to contradict the premise of movies like *Up* and others that have some kind of older person who espouses a "get off my lawn" mentality.

JLR: Ha, yes. But the bias in this case is about memory and refers to the types of episodes we're more apt to remember once we get into our seventh and eighth decades of life.

RB: Ah, so you're saying that as people age they are more likely to remember the positive experiences of their lives than the negative ones?

JLR: Yes, so, if any of our readers are still lamenting what Sally Parkin[1] said in the 11th grade in front of the WHOLE honors trigonometry class, they needn't worry that the memory will haunt them forever.

RB: That is a disturbingly specific example.

JLR: Yes, but luckily in a few decades, it won't matter anymore. Oh. And as for the "get off my lawn" vibe, I don't think the positivity bias applies to that. And based on personal experience, I think that may kick in sooner rather than later.

RB: There is way too much to unpack here. Let's just start discussing the different types of memory.

---

[1] Not her real name.

© Jerome L. Rekart and Rebecca Baker 2025
J. L. Rekart and R. Baker, *Designing for Human Intelligence in an Artificial Intelligence World*,
https://doi.org/10.1007/979-8-8688-1418-1_9

CHAPTER 9   I REMEMBER WHEN...OR DO I?

JLR: Agree, just give me a second; it looks like a Frisbee just landed on my porch.

Memories Come in Different Flavors...

Ask the average person about their oldest *memory* and chances are good that they will regale you with a personal experience from when they were between two and a half and four years old. They may tell you about a birthday, being pushed on a swing, obtaining a pet, seeing fireworks, or something else (hopefully) that brought them joy. It is likely that the memory they recall feels like it has been stored perfectly in their mind. Indeed, as you're reading this, you may be recollecting your own past and thinking back to when your age was measured in single rather than double (most likely) or triple (you go!) digits.

Except, they (you) wouldn't be entirely correct.

Though the recalled personal experience that we remember as the first in our personal chronology may be our oldest *episodic* memory, this is not our oldest memory *per se*. To understand, this we should first define what is meant by a "memory." **A memory is merely information that has been stored within our brains**. This information can be a skill, fact, concept, or an image of a fancy hippopotamus wearing a top hat. Given this definition, then all of the skills and knowledge we have accumulated across our lifespan must be considered when thinking about what one's oldest memory could be.

CHAPTER 9   I REMEMBER WHEN...OR DO I?

Given that definition, then in actuality, our oldest memories are those that relate to

- Crawling, climbing, and walking
- Eating/using utensils
- Understanding that some combination of "m" sounds equate to a female parent or caregiver and/or
- That some combination of "b," "p," or "d" sounds refer to a male parent or caregiver, and
- That the sounds associated with our own names are a means of identifying ourselves

211

CHAPTER 9   I REMEMBER WHEN…OR DO I?

The first two bullets from the list (above) identify skills and physical feats that we accomplish throughout our lives – whether using chopsticks or playing piano (or playing "Chopsticks" on the piano) – and are referred to as *procedural* memories. The other bullets, which contain answers to questions that begin with *who, what, where, when, why*, and sometimes *how*, are called *semantic* memories.[2]

There are several reasons why we differentiate between different memory types. First of all, it is because they are actually dissociable – meaning that we can have damage or loss of one type that doesn't impact others. This dissociation is seen in extreme cases of Alzheimer's disease, which in the early and mid-stages selectively impacts episodic and semantic memories while leaving procedural memories intact.[3]

Second, different memory types are encoded with varying levels of specificity and persistence. Whereas episodic memories have the potential to be modified and changed every time we remember a given moment from our lives, semantic memories are much more resistant to change.[4] For this reason, researchers know that episodic memories are the ones that are least likely to be accurate. In fact, many episodic memories are actually reconstructed, meaning that although we likely store the gist of what happened to us at a given moment, like a birthday party (yay!) or work event (uh…sure…woohoo), we do not store the information exactly as it occurred. When it comes to episodic memory, we should not think of our brains as digital video recorders but rather as note-takers who madly

---

[2] Squire, L. R., & Zola, S. M. (1996). Structure and function of declarative and nondeclarative memory systems. *Proceedings of the National Academy of Sciences, 93*(24), 13515–13522.

[3] De Wit, L., Marsiske, M., O'Shea, D., Kessels, R. P., Kurasz, A. M., DeFeis, B., … & Smith, G. E. (2021). Procedural learning in individuals with amnestic mild cognitive impairment and Alzheimer's dementia: a systematic review and meta-analysis. *Neuropsychology review, 31*, 103–114.

[4] Scully, I. D., Napper, L. E., & Hupbach, A. (2017). Does reactivation trigger episodic memory change? A meta-analysis. *Neurobiology of learning and memory, 142*, 99–107.

CHAPTER 9   I REMEMBER WHEN...OR DO I?

try to capture as much pertinent information as possible so that it can then be pieced back together later to form a coherent (or not) idea of what occurred. These mental note-takers don't write about the background, clothing, etc. in great detail but instead will jot down generalizations like "birthday," "winter," or "spaghetti dinner." These generalizations are then used as rough cues to reconstruct our memories when we relive them. Thus, word-for-word conversations, specific details about what people may have been actually wearing, room decor, the full menu of a meal, etc. may be (and are likely) completely lost.

And yet, when we remember that work event, we have the experience of remembering something as though we *did* store all of the details, even though we didn't. This is the reconstructive component. Our brains fill in the gaps that don't exist in memory by making assumptions about what things could have looked like by using those general cues, what conversations were likely to sound like given what we know about the folks who were there, etc. This is not unlike what we described for language (i.e., the hedging of bets using statistical frequencies) and is surprisingly similar to how artificial intelligence sometimes hallucinates (i.e., by filling in knowledge gaps with statistically likely events).

Unlike episodic memories, semantic memories are much more resistant to reconstructive contamination (i.e., the process of modifying memories every time we recall them). This is in part because there are fewer details to associate with the event. Thus, if you were feeling like the last few paragraphs indicated that we don't remember anything, you can take solace in the fact that we extract almost all of our semantic memories – which, again, are much more resistant to corruption and reconstruction – from the episodes in which they were formed. If that last sentence didn't completely jive, then the next example will hopefully provide some illumination.

213

CHAPTER 9   I REMEMBER WHEN…OR DO I?

Take a moment and think about the US Government (don't worry, we're not getting political with this). OK, so here is our question for you: How many branches of government are there, and what are the names of those branches?

[cue music from *Jeopardy*]

Got it? Great. Most students in US-based public schools learn the three branches of government between the third and fifth grade. So, if you were able to recall the names of all three (and were educated in the States), then chances are good that the very first time you learned them you were sitting in a classroom filled with desks and your teacher was going over these facts with you. They may have been reading from a text or conducting a Socratic seminar or standing at a black/whiteboard and writing out the answers. Regardless of how the information was presented or even when, if you remember it, then you must have learned it at some point, even if that only involved you alone in a room reading a passage in a textbook.

Here is where the distinction between semantic memory (the name of the three branches) and episodic (the memory of actually learning the three branches) should become apparent.

- Do you remember sitting in that classroom? Can you remember the words on the overhead transparency, worksheet, blackboard, whiteboard, or in the text?

- Do you have any recollection of the day that Mr(s). X, your third, fourth, or fifth grade teacher, first introduced the ideas of the legislative, executive, and judicial branches?

If you're being honest, then chances are excellent – like outstandingly, incredibly almost 100% true – that even though you have an intact semantic memory of what those branches are (i.e., you correctly identified executive, legislative, and judicial), you have absolutely no accompanying episodic memory of when or how you actually learned them. Everything else about the moment in time those concepts were introduced, including

214

CHAPTER 9   I REMEMBER WHEN...OR DO I?

what your teacher was wearing, what your classmates were doing or saying during that part of the lesson, how the class was decorated, what you had for lunch that day, the color of your folder, pencil, or the cover of the textbook, is all lost.

This preferential storage of semantic memories and loss of episodic ones is true for the overwhelming majority of facts that we accumulate not only throughout school but our lifetimes.

And yet, even for semantic memories, not everything we think we've learned stays in our memories.

Going back to facts learned by most third to fifth graders, we'll now ask a few that wouldn't depend on being in elementary school in the United States. Globally, most students learn various facts about the human body during the educational interval that occurs between 9 and 11 years of age (i.e., third to fifth grade in the United States). Now we're going to see how many of the following facts you can answer just by relying on your memory (nobody else is around, so don't cheat and look anything up on the Internet):

1. What is the largest organ of the human body?

2. What is/are the largest bone(s)?

3. What is/are the smallest bone(s)?

4. What is the purpose of the lymphatic system?

Were you able to answer all four?

Again, chances are excellent that you (1) learned all of this information at one time (or at least were presented with it) and (2) that if you're not in the medical or biological fields (or a trivia guru) that you struggled with at least one of these (the answers, by the way, are in the footnotes to this chapter[5]).

---

[5] (1) Skin; (2) femur; (3) stapes (stirrup), incus (anvil), and malleus (hammer; in the inner ear); (4) often called the "sewer" of the body, the lymphatic system is responsible for maintaining fluid balance and facilitating immune responses.

215

CHAPTER 9 I REMEMBER WHEN…OR DO I?

All that was asked this time was for you to recall facts, which would have been stored in semantic memory, yet, if you couldn't recall one or more (or did so incorrectly), then that means that you in fact do not have a memory for that information. So what happened? Read on…

…And Have Different Expiration Dates

As the little quiz in the previous example made clear, we can know things at one point in time and yet forget them later on. Indeed, we have all experienced the phenomenon of not being able to recall an answer that we swear we know. And yet, there is other information that we may want to forget but can't because we have little use for it. Thus, some information lasts a lifetime and other bits and pieces much less. The reason for this is that just as we have different categories of memories based on the *type* of information to be stored (e.g., experiences, facts, skills, etc.), we also have different memories based on *how long* they are stored.

If we go back to the definition of memory as information that is stored within our brains, then there are some forms which are so ephemeral that we may not even realize that they are actually memories. And then of course, there are those that last – hopefully – as long as we do, like our knowledge of our own name, what we like and dislike, and the playback of cherished experiences with our loved ones.[6]

Traditionally (or at least dating back to the late 1960s), memory has been divided into three temporal categories: sensory memory, short-term (or working) memory, and long-term memory.[7] Each memory is stored for a limited (sensory, short-term) or undetermined (long-term) period of

---

[6] Even though it is likely that what we are actually recalling is just a reconstruction of an event and not the actual occurrence itself.

[7] Atkinson, R., & Shiffrin, R. (1968). Human memory: A proposed system and its control processes. In K. Spence & J. Spence (Eds.), The psychology of learning and motivation (Vol. 2, pp. 89–195). New York: Academic.

216

CHAPTER 9   I REMEMBER WHEN…OR DO I?

time in what psychologists call an **engram,**[8] **which is the actual, physical representation of information within our brains**. The persistence of memory is caused by duplication (and reduplication) of the engram into various regions that underlie the progressively longer-lasting forms of memory (i.e., sensory memory gives rise to short-term, which in turns gives rise to long-term).

The two most widely understood forms of sensory memory are related to the processing of information that comes through our ears and eyes. These forms, which are referred to as echoic and iconic memories, respectively, are incredibly short-lived but useful. In order to understand echoic memory, we first have to recall that our sense of hearing is caused by a transfer of the vibration of air against our eardrums into neural signals (i.e., release of neurotransmission). The transduction of the sound signal takes place in our inner ears (in the cochlea) when microscopic hair cells are bent. The act of bending the hairlike extensions of these cells allows us to not only hear but actually differentiate between sounds, like middle C and D on a piano. The physical bending of the cells actually lasts longer than the sound that caused the bending, which means that we actually experience the sound for a slightly longer period than we otherwise would. In the absence of any additional sounds (which unbend or re-bend the cells), a given sound can persist for anywhere from one to three seconds. Thus, the information "echoes" in a sense and persists within our brains: thus echoic memory.

Within our eyes, there are chemical reactions within our rods and cones (at the back of our retina) that transform light energy from outside our bodies into a message that our brains understand, which again is release of neurotransmitters. Iconic memories do not last nearly as long as echoic ones, with a persistence of visual images for about 200 milliseconds (or one-fifth of a second) beyond when we *actually* saw something. We can

---

[8] Lashley, K. (1950). In search of the engram. In *Symposia. Society of Experimental Biology* (Vol. 4, pp. 454–482).

217

CHAPTER 9   I REMEMBER WHEN...OR DO I?

sometimes experience this phenomenon when we look at something really brightly colored and then quickly change our gaze to a uniformly white or black wall and see a ghostly or phantasmal blob or outline of what we were previously looking at. This process resets itself much faster because rather than a physical process (i.e., microscopic hairs bending and unbending), the chemicals within the cells merely need to be reset at a molecular level. This information, though fleeting in duration, is actually massive within its scope. This means that the entirety of whatever was heard or seen is stored for the period of time it persists within either of the two sensory memory stores.

If the information is deemed important (see Chapter 6), it enters into our short-term memory store. Although it lasts longer than sensory memory (about 30 seconds if you aren't distracted by new information), short-term memory has a severe limitation on the quantity of information that can be stored for this period of time.

What kind of limitation? Let's see for ourselves. First, get yourself a blank piece of paper or open a note app on a mobile device so that you can test your own memory. Once that is ready (and after you've finished reading this paragraph so you know the full extent of the instructions), you will read the next paragraph which contains a list of unrelated words. What you'll want to do is read all of the words in the paragraph, and when you've finished (and without peeking), then write down as many of the words that you can remember on the piece of paper or type them into the app. Once you've exhausted your memory, then read the passage after the list to see what the results will show us (or, if you're not of the audience participation type, just skip ahead now).

Ready? The list appears in the next line:

Dog bud top hug rot apt cut jig map hew sad pop bag yew.

CHAPTER 9   I REMEMBER WHEN...OR DO I?

Now, what you'll want to do is compare your list of words against the actual list (above) and count the words that you correctly remembered. Even though we don't know you (but we are so appreciative that you are reading our book), we feel confident that you remembered somewhere between five and nine of those words. Why is this? Well, in 1956, George Miller published a seminal paper showing that the capacity of a human's short-term memory span is seven plus or minus two items (which gives us the range from five to nine).[9] And the longer you make someone wait to write down the words (particularly if longer than about half a minute), the fewer they'll remember – particularly if you distract them with some other task that requires thinking (like counting backwards from 29 by 3s).

Though first elaborated decades ago, we now know that this limit represents the upper range of what humans can remember immediately after information is presented (so don't feel bad if you remember fewer than five). We can demonstrate this as well using the same setup as previously, but now with a different set of words. So again, ready that paper or get the note app ready.

Let's try again with this new list (below):

Healthy genuine officer arrange typical chapter serving interim filling maximum quarter produce nursing evening.

Now, how many words did you remember? Chances are excellent that you remembered fewer than you did with the first list even though both contain the exact same number of words (14). What you may have noticed is that the words in the second list were longer than those in the first (seven letters rather than three), which actually increased the total amount of information that you were processing (think sounds). The net result: you probably remembered three to five words total.

---

[9] Miller, G. A. (1956). The magical number seven, plus or minus two: Some limits on our capacity for processing information. *Psychological review*, 63(2), 81–97.

CHAPTER 9   I REMEMBER WHEN...OR DO I?

This second demonstration reinforces that not all information is the same. The more complex the information, the less of it we can *passively* hold on to in the short term. Did you note that we italicized the word "passively" in the last sentence? That is because there is something we can do to expand the capacity of our short-term memory: we can expend cognitive resources and *work* with the information.

Working Memory: Blood, Sweat, Tears, and...Neurotransmission

It is probably apparent that being able to transform our short-term memory into a *working memory* store is useful as it allows us to transcend the typical 30-second lifespan of most short-term memories and facilitates cramming more than 5 or 7 or 9 bits of information into our brains. And how do we perform this *work*? There are two main ways to do so: both of which rely on the neural machinery responsible for our senses of sight and hearing.

Think back for a moment to any of the images you've encountered in this book. See if you can visualize any of the ones listed below or others from the text. Images such as

- A many-armed multitasker flanked by a unicorn and a yeti

- A rendering of a certain famous Jedi master getting a piggyback ride from his apprentice in a swamp

- A child's rendering of an astronaut being ridden by a horse

If you were able to successfully remember any of those images from the book, this was facilitated by your mind finding the engram for that image in your long-term memory and bringing it forward for you to "see" using

220

CHAPTER 9   I REMEMBER WHEN...OR DO I?

your mind's eye. This act of imagination used the visual machinery of your occipital cortex and gave you the impression that you were "seeing" the image inside your mind. This form of memory "work" uses the so-called "visuospatial sketchpad" (or scratchpad, as it is sometimes referred) and is activated not only when we remember something visual but also when we try to keep an image we've just encountered in our short-term memory (for longer than 30 seconds).

We can also keep auditory information active in our working memory using a similar process for sounds. Thus, by activating our "phonological loop," we use our mind's "voice" (and "ear") to rehearse information that we want to hold on to. For example, let's say that you are in a downtown urban area and realize that you don't have your phone (rest easy – this is just a hypothetical – no need to panic), but you just witnessed one car careen into another and drive off! Being the responsible person that you are, you want to make note of the license plate of the offending vehicle that is now speeding away with no regard for the damage it has caused. With no phone camera to take a picture, you are left with no choice but to try and remember the license plate until you can write it down and pass it along to the police.

CHAPTER 9   I REMEMBER WHEN...OR DO I?

How will you remember this seemingly random series of letters and numbers and fulfill your duty as an upstanding citizen? Most likely you would try to "hold" on to it by saying it over and over again in your head. Thus, by using the neural machinery that processes sounds, you'd be able to hold on to the information for as long as you need to, provided you kept rehearsing it.

So by using both the visuospatial scratchpad (i.e., our visual imagination) and the phonological loop (i.e., our mind's internal dialogue), we are able to keep information activated in our short-term memory stores indefinitely. By acting on new information in this way, we increase the likelihood that engrams representing it will be created in our long-term memory stores as well, thus giving us access to the information whenever we want it. This dissociation between working/short-term memory and long-term memory allows us to deal with the real-life practicalities that there are some forms of information we need for longer than 30 seconds but do not need for the rest of our lives (like directions somewhere we'll only visit once, menu items at a restaurant, license plates of irresponsible drivers, etc.).

We can also exploit the fact that as we age, there are fewer and fewer truly "new" things that we encounter. This means that we can use some tricks to use past memories to facilitate our ability to hold onto a great deal of newer information. Let's take a look at how this is done by having another list for you to memorize.

The list we're going to use this time will have numbers rather than words, but the same rules apply: no peeking and only read through the list once before trying to recount (pun intended!) what you saw.

Ready?

2 0 0 1 1 2 1 5 1 9 6 9 1 9 4 5

At first glance, this might seem unnecessarily tricky and perhaps even cruel, seeing as there are 16 bits of information (i.e., digits), which clearly flies in the face of the 5–9 limit we discussed earlier in this chapter. That is unless you recognized a pattern found within those 16 digits. If you noticed

CHAPTER 9   I REMEMBER WHEN…OR DO I?

the pattern, then for you these wouldn't be just any 16 random numbers, but dates, with each group of four numbers representing a particular one of note.[10] In this way, folks are able to use past information (i.e., knowing that groups of four digits can represent dates) to remember complex pieces of information.

This technique, called "chunking,"[11] is in fact not only useful for memorizing large groupings of numbers but is in fact the foundation upon which our ability to read rests.[12] This type of work occurs in working memory, which is seen by some researchers (including Dr. Rekart) as THE seat of consciousness.[13]

Let's take a moment for that to sink in. Essentially, all of our moment-to-moment experiences, likes, dislikes, ongoing internal dialogue, etc. operate within our working memory stores (both the visuospatial sketchpad and the phonological loop) – meaning the location within our mind that processes our current state, makes sense of it vis-á-vis the past (i.e., compares with our long-term memories), and allows us to make decisions is largely where we "live" when we're awake. If we want to use a computer analogy, we could think of working memory as our desktop, which relies on RAM, and it is only when we really want to keep something that we store it in our hard drive (i.e., long-term memory).

This process of duplicating information that is in short-term/working memory stores into our long-term memories can unfortunately be disrupted. The hippocampus, a seahorse-shaped structure within our brains, is responsible for this process, and injuries to it cause severe

---

[10] The first is an Arthur C. Clarke reference, the second is when the Magna Carta was signed, the third when a human walked on the surface of the moon, and the last one marked the end of WWII.

[11] Mathy, F., & Feldman, J. (2012). What's magic about magic numbers? Chunking and data compression in short-term memory. *Cognition, 122*(3), 346–362.

[12] Jeffries, S., & Everatt, J. (2004). Working memory: Its role in dyslexia and other specific learning difficulties. *Dyslexia, 10*(3), 196–214.

[13] Baddeley, A. (1993). Working memory and conscious awareness. In *Theories of Memory* (pp. 11–28). Lawrence Erlbaum Associates.

CHAPTER 9   I REMEMBER WHEN...OR DO I?

impairments.[14] Indeed, the hippocampus is where debilitating disorders like Alzheimer's disease first manifest, which is why individuals first start to have issues remembering what they were doing in a given moment or were thinking about (i.e., short-term memory loss; also called anterograde amnesia). We spend most of our lives taking for granted how important this can be, that is, until we place a pot on the stove and forget that it was ever lit! As diseases like Alzheimer's progress, the ability to store any new episodic or semantic memories goes from being compromised to completely impossible. This phenomenon is depicted quite well in the film *Memento* by Christopher Nolan (which is not about Alzheimer's disease but how hippocampal damage would actually present).[15]

Making Memorable Designs

Although we could easily go on with the philosophical and existential ramifications of a memory store as the engine of sentience, we'll bring it back to how this information can be put to practical use.

Create Familiar Experiences

Your users are comparing everything they interact with against what they already know or anticipate to be true. Though you may not think of it as a "memory," familiarity is in fact exactly what this is. Studies have shown that familiarity – for example, with website design, layout, and navigability –

---

[14] Squire, L. R., & Zola-Morgan, S. (1991). The medial temporal lobe memory system. *Science, 253*(5026), 1380–1386.

[15] Not only does it do a top-notch job of representing the dissociation of different temporal memory stores (i.e., long-term vs. short-term), it is also a great, stylistic thriller.

CHAPTER 9   I REMEMBER WHEN...OR DO I?

leads to trust (which, in turn, positively impacts purchasing).[16] Use established patterns to help users be efficient rather than making them learn novel interactions.

Keep It Simple

Keep in mind that cognitive resources are limited by working memory constraints. This means that whenever you can reduce the load by simplifying things for your users, you should. In so doing, you will free them to make decisions based on the information that you want them to use (i.e., navigation aids, marketing materials, special offers) rather than other processes that can be more chaotic and, thus, less predictable.[17] We will discuss how design can influence decision-making in the next chapter. Clean and simple design supports a good user experience. The more distractions you have on your site, the more cognitive work you are creating for your user. For an example of how NOT to design your website, consider the infamous Ling's Cars[18]:

---

[16] Kaya, B., Behravesh, E., Abubakar, A. M., Kaya, O. S., & Orús, C. (2019). The moderating role of website familiarity in the relationships between e-service quality, e-satisfaction and e-loyalty. *Journal of Internet Commerce, 18*(4), 369–394.

[17] Sicilia, M., & Ruiz, S. (2010). The effects of the amount of information on cognitive responses in online purchasing tasks. *Electronic Commerce Research and Applications, 9*(2), 183–191.

[18] https://www.lingscars.com/order

CHAPTER 9   I REMEMBER WHEN...OR DO I?

Embrace Chunking

Don't treat your designs like problems to be solved from the *The DaVinci Code*. Group together similar items, tasks, etc., so that their likeness is obvious. For example, if I wanted you to see the four dates in the last example, you should make sure that the spacing between the four-digit sequences was much greater than the total spacing (e.g., "1969 1945" not "1 9 6 9 1 9 4 5"). By doing this, you make it apparent what patterns can be used to "chunk" information, which will free users' working memory to process other things. Remember George Miller's research? Those findings led Bell Labs to design phone numbers to be seven digits long, broken into two chunks of three and four digits.

CHAPTER 9   I REMEMBER WHEN…OR DO I?

Cognitive Cartography (and Other Alliterative Matters)

If we go back to the metaphor of a brain as a computer, then long-term memory is akin to a hard drive. Indeed, though this may have once been true and definitely fits with the notion of storage "on prem,"[19] it may be more correct to think of it instead as unlimited storage in the cloud (which stylistically is a nice throwback to representations of thought in comics). This is because although there are workarounds to the limitations of working memory, those checks will always be there. With long-term memory, on the other hand, researchers have yet to find the limits of human storage capacity. Thus, it is hypothetically possible that one's other organs will give out long before ever reaching the limits of mnemonic capabilities.

Knowing that there aren't limits to what *can be* learned doesn't make acquiring new information any easier, however. That is until one realizes that we can effectively gain new knowledge and design in a way that makes it easy for others to do so by exploiting the same serendipitous evolutionary step forward that facilitated our species' ability to learn a myriad of new things in the first place.

We have already identified the hippocampus as the brain structure responsible for transferring information into long-term memory, but if we rewind the evolutionary clock quite some time (give or take a few hundred thousand or million years), then we would see that its original function was not to remember birthdays, the definition of the word sesquipedalian, or that the Chicago Cubs were founded in 1869. No, the original – and continued – function of the hippocampus was to help us effectively navigate our world.

---

[19] For those not used to thinking about data storage, "on prem" refers to data storage that physically takes place at the same location where it is created/used rather than in "the cloud."

CHAPTER 9   I REMEMBER WHEN...OR DO I?

In order to explore these concepts, let's rewind the clock. We won't go as far back as the divergence of humans from other primates that occurred about 6 million years ago but will rewind to around 45,000–60,000 years in the past. At this point in time, it wasn't just our human ancestors (the Cro-Magnon) who walked on two legs around various locales on earth but also two other early humanlike hominid species: the Neanderthals and the Denisovans.[20] We're choosing this point in time, because having one or two other hominids to worry about (depending on the location) would have made life very interesting – and by interesting we mean SCARY – for our ancestors. Thus, in addition to worrying about disease, unpredictable weather patterns, scarcity of food, etc., our ancestors also had to navigate a world where other physically similar but identifiably different species hunted, gathered, ate, procreated, and lived.

And those activities are key because they all involve keeping track of where one can find necessary items (like food, a cozy cave, or a mate) and where one should steer clear of (the field frequented by the big grumpy fellow with the highly pronounced brow, the place where the bush was on fire after the storm, etc.). All of this information, as it is for other species, is entered into long-term memory via the hippocampus and would have been at this point in time for our direct ancestors (the Cro-Magnon) as well as our distant cousins (the other two).

But our ancestors didn't stop at just remembering places. They (and the other early hominids[21]) also started to use tools, form complex social hierarchies, travel great distances, and produce arts and practical crafts. These processes could only have been facilitated by species that were able to take in not only what is found in a place (spatial information) as well as what

---

[20] Bergström, A., Stringer, C., Hajdinjak, M., Scerri, E. M., & Skoglund, P. (2021). Origins of modern human ancestry. *Nature, 590*(7845), 229–237.

[21] Hardy, B. L., Moncel, M. H., Kerfant, C., Lebon, M., Bellot-Gurlet, L., & Mélard, N. (2020). Direct evidence of Neanderthal fibre technology and its cognitive and behavioral implications. *Scientific Reports, 10*(1), 4889.

CHAPTER 9   I REMEMBER WHEN...OR DO I?

occurs there over time (temporal components). This latter process in essence piggybacked on to our ability to detect and encode (i.e., make sure we later remember) where we are in three-dimensional space relative to geographic markers (i.e., landmarks) and used it to make sure we remember a series of steps (first you do X, then you do Y), as would be required to manipulate textiles, usher in the advent of agriculture, etc. But what are we to make of the fact that this oddly named part of the brain is responsible for our ability to remember places and, well, everything else that is fact- or event-based?

To answer this question, we're going to time travel one last time. We won't return entirely to the present day and place (wherever you may be) but will stop instead in 500 BCE in the Thessaly region of Northern Greece. We are stopping here because legend has it (as relayed by Cicero[22]) that the Greek poet, Simonides of Ceo, was dining at a banquet for a wealthy nobleman named Scopas. Being a poet, Simonides recited some lines once dinner was over, and then, when finished, he was called outside to meet with some messengers who arrived to speak with him (showing, as we'll see, some suspiciously good timing). While he was outside talking to the messengers, calamity struck, and the roof of the banquet hall and supporting walls all collapsed, crushing everyone who was left in the hall and killing them (bet you didn't see that coming...alas, they most likely didn't either).

The scope of the tragedy was so extensive that officials were unsure how they would identify the remains of individuals to ensure a proper burial. Luckily (for many reasons), Simonides, as the only survivor, found that by imagining where he was in the room during the dinner, he could identify where various individuals were sitting in the room. This technique, which he talked about and used later, is sometimes called a "Memory Palace" or, more commonly (and with less grandeur), the Method of Loci (MoL).[23]

---

[22] Mankin, D. (Ed.). (2011). *[De oratore, book III]; Cicero, de oratore, book III*. Cambridge University Press.

[23] Rethinking memory as a feature of mapmaking | Jerome Rekart | TEDxSNHU.

CHAPTER 9   I REMEMBER WHEN…OR DO I?

This methodology, which involves thinking about whatever is to be remembered in a known geographic location, has been shown time and time again to be one of, if not the, most effective mnemonic techniques that can be used.[24] Now, in the present day (we're back from our trip – wow we went far and wide!), we can utilize the MoL to remember almost anything. Thus, for our own purposes (just day-to-day), we can take any list of things that we need to remember and visualize them at various locations in a well-trod location, like our local grocery store or even inside our refrigerator. In the former example, we would imagine the items to be remembered on shelves or in various aisles that we have been traveling for some time and know quite well. In the latter example, we see the utility of this technique is that we can think of almost anywhere we know as a map.

In this context, the contents of our cabinets are stored by our hippocampus no differently than trees along a trail. That bottle of catsup (or is it ketchup) and a tall oak tree are both treated as landmarks in a place. Thus, we can just easily "place" (using our mind's eye) concepts, objects, or items to be remembered among our plates, cups, saucers, and boxes of breakfast cereal.

Designing with Memory Maps in Mind

And all of this is related (surprise!) to our discussion of storytelling as a design element in the last chapter. Indeed, maps have been seen as storytelling vehicles (and vice versa) for years.[25] After all, what is a map but a series of points along a story and what is a story but a collection of events

---

[24] Wagner, I. C., Konrad, B. N., Schuster, P., Weisig, S., Repantis, D., Ohla, K., … & Dresler, M. (2021). Durable memories and efficient neural coding through mnemonic training using the method of loci. *Science advances*, *7*(10), eabc7606.

[25] Roth, R. E. (2021). Cartographic design as visual storytelling: synthesis and review of map-based narratives, genres, and tropes. *The Cartographic Journal*, *58*(1), 83–114.

CHAPTER 9   I REMEMBER WHEN…OR DO I?

that occur through time and/or space. So, other than being incredibly useful for all of us as humans, there are also some practical uses of this natural proclivity toward and for maps.

Structuring Flows

Multiple studies have shown that users prefer to navigate through files on their computer rather than use a search tool (even though the latter is more efficient). This preference has been shown to be a byproduct of hippocampal activation that mimics what happens when users look through real (i.e., physical) files and locations.[26] Thus, whenever possible, try to mimic processes that "feel real," and follow processes and steps that one might encounter outside digital realms.

This preference is particularly relevant to document design. Visualize a physical book you've read recently. Recall a passage in that book that was particularly striking for you. Chances are you could point to where in the book that passage was (about ⅓ of the way in, on the right side, about half way down the page). When designing virtual experiences, particularly virtual reading experiences, we need to remember that while the Search function is efficient, it is not effective when you include a human element. We want (perhaps need) a sense of where something is in our virtual physical space. Tables of contents, structured document design with headings, and similar locations are all useful in improving user recall and satisfaction in the results. Think of a user or customer journey as a story with physical elements. Those elements are most likely not *actually* physical but will be treated by your users' brains as such when they navigate through the spaces you design.

---

[26] Benn, Y., Bergman, O., Glazer, L., Arent, P., Wilkinson, I. D., Varley, R., & Whittaker, S. (2015). Navigating through digital folders uses the same brain structures as real world navigation. *Scientific reports*, 5(1), 14719.

CHAPTER 9   I REMEMBER WHEN...OR DO I?

OLD DOESN'T MEAN BROKE

Early in my career, I worked as a technical writer and technical writing manager. In the beginning, when documentation moved online, we mimicked the paper experience down to having a "this page intentionally left blank" page when appropriate. As technology progressed, we tried to rethink how to structure documents to take advantage of the efficiencies being offered. However, we consistently got push back from users who wanted some forms of the old structures. Initially, we saw this as reluctance to adopt new ideas (remember, humans don't like change!). Over time, it became clear that it was more than that – some structures were useful for navigating and visualizing content in a way that a flat search could never mimic. Some structure fell by the wayside (bye bye index!), but others like the table of contents stayed as a map of the document contents, albeit linked and much more useful. As I moved into design, many of these lessons held.

Mimic Tangibility

Although we evolved to use our hippocampi for many things, we have not evolved (nor will we without any selective pressure) to be completely virtual. To this end, whenever you can pair stimuli so that physical elements are mimicked (think of the clicks that accompany virtual dials or knobs), you will reduce cognitive loads on working memory and activate innate navigational mechanisms that will help your users focus on what you want them to focus on (which isn't getting lost or figuring out what "this" does). This understanding is what makes skeuomorphic design effective. Virtual folders that look like paper folders, virtual buttons that appear to be depressed when clicked on (and make a click noise), animations that mimic a page flipping – all of these improve our ability to work more effectively with virtual objects.

232

CHAPTER 9   I REMEMBER WHEN...OR DO I?

Remembering Some Cognitive Biases

We see things not as they are but as we are.

—Anaïs Nin

Remember cognitive biases from Chapter 3 (see what we did there)? There are a whole host of cognitive memory biases (unsurprising given what we've covered up to this point). Let's look at a few:

Rosy Retrospection. We tend to remember the past as having been better than it really was – the "good old days" or "glory days." Because of this, we may view past trends and events as more favorable than our current situation. Products or messages that use nostalgic representations of the past (Classic Coke, 90's edition 501 jeans) appeal to this bias and can be used to get us to make choices based on our positively biased memory rather than an evaluation of the worth of the current product. This is related to, though different than, the positivity bias that was discussed

CHAPTER 9   I REMEMBER WHEN…OR DO I?

in the opening dialogue of this chapter, though both are believed to be caused by age-related shifts from remembering dangers (i.e., negative stimuli) to focusing on emotional regulation.[27]

Consistency. We remember ourselves inaccurately. With consistency bias, we remember our past selves as being consistent with our present selves. We gloss over the often deeply significant changes any human goes through as they age. Related to this is choice-supportive bias in which we remember the choices we made in the past as being better than the alternatives.

Hindsight. "Hindsight is 20-20." Even though we were surprised at the time, we remember incidents in the past as being better than they were. We recall our level of understanding and prediction as much higher than it actually was as well.

Availability. If we can easily remember something, we believe that thing is more typical or usual than things we have difficulty recalling. This is also why we are bad at making logical decisions (see Chapter 10 for more on that). Have you ever noticed strange things happen more frequently during a full moon? In 2011, a study[28] looking into this phenomenon found that while 40% of medical professionals in emergency rooms believed that there was an increase in blood loss and emergency frequency during the full moon (and Friday the 13th), when looking at the data, it turns out that there was no increase at all. Study after study[29] show no correlation between the lunar phase and emergencies or strange occurrences. So how is it that so many people have a belief that contradicts the facts?

---

[27] Reed, A. E., Chan, L., & Mikels, J. A. (2014). Meta-analysis of the age-related positivity effect: age differences in preferences for positive over negative information. Psychology and aging, 29(1), 1–15.

[28] Schuld J, Slotta JE, Schuld S, Kollmar O, Schilling MK, Richter S. Popular belief meets surgical reality: impact of lunar phases, Friday the 13th and zodiac signs on emergency operations and intraoperative blood loss. World J Surg. 2011 Sep;35(9):1945-9.

[29] Thompson DA, Adams SL. The full moon and ED patient volumes: unearthing a myth. Am J Emerg Med. 1996 Mar;14(2):161-4.

CHAPTER 9   I REMEMBER WHEN...OR DO I?

Consider you are out with your friends listening to an amazing classical guitarist at your favorite cafe while you sip a hot cup of coffee, letting the bitter taste rest briefly on your tongue before letting it warm you from the inside. A man walks into the cafe wearing an outrageous suit of lime green covered in blinking lights. You glance outside the window and notice a full moon and think "oh yeah, full moon the crazies are out!" Two weeks later, you are at an Italian restaurant with a close friend listening to the strains of Frank Sinatra while you enjoy the savory taste of garlic bread. A woman walks in wearing the most outrageous hat you've ever seen – a large pink tower of tulle crested with what looks to be a plush flamingo, complete with a flashing "Viva Las Vegas" mural. You glance out the window, but no moon lights the night, and you think "Wow, that's pretty crazy." The added detail of the full moon creates a detail we can easily recall when asked "Do crazy things happen during the full moon?" that easily comes to mind.

235

CHAPTER 9   I REMEMBER WHEN…OR DO I?

Recognition over Recall

Our ability to remember information that has been presented in a linear order (think grocery lists) is a well-documented phenomenon. This phenomenon, the serial-position effect, holds that our memory for the first items of the list will be the ones that are remembered best in the long term (i.e., the primacy effect) and those at the end of the list will be remembered well right after we get done reading or listening to it but will gradually disappear into the ether with the others from the middle (i.e., the recency effect[30]).

For example, let's imagine that we were asked to get the following items and didn't have the opportunity to write it down or somehow make note of the list in our ever-ubiquitous phones (though not in this instance):

- Butter

- Carrots

- Lima beans

- Paper towels

- Celery

- Diced tomatoes

- Ground turkey

- Chicken broth

- Masa

- Parsley

- Onions

- Cardamom

---

[30] Feigenbaum, E. A., & Simon, H. A. (1962). A theory of the serial position effect. *British Journal of Psychology*, *53*(3), 307–320.

CHAPTER 9   I REMEMBER WHEN…OR DO I?

This means – assuming that we didn't know how (or if) the ingredients go together – we would have the greatest likelihood of buying butter, carrots, lima beans, and paper towels (the first few items from the list); the next greatest likelihood of getting the masa, parsley, onions, and cardamom; and would most likely not get the diced tomatoes, ground turkey, or chicken broth. The reason for this effect is that our brains have enough time to process the first items into working memory and then reduplicate in long-term memory. While they are being consolidated in long-term memory, our brains are also trying to "hold onto" the new information in working memory, which would be the last few items. The reason why the recency effect isn't as efficient at encoding information into long-term memory is because of those pesky limitations on working memory that we discussed earlier. Thus, if your designs in any way incorporate lists or series of steps or items, be sure to accommodate for serial position effects. Indeed, these effects have been shown to have robust impact on marketing, with television commercials remembered better when they are the first in a grouping of televised ads rather than the last few.[31]

Now, in the last example, our presence in a grocery store or shopping center would help us to remember what to get due to cues and "jogs" to our memory from all of the items on the shelves. This is because of the difference between recognizing previously presented information and recalling it.

In our experience as educators and learners, we have found that most students prefer to have their understanding evaluated by multiple-choice tests rather than essays. Chances are good that many of you feel or felt the same way when you were in school. Though there are several reasons for this, not the least of which is that one of those methods provides the student with the possibility of getting an answer correct via guessing, it is in part because the act of recognizing information you've seen previously (which

---

[31] Terry, W. S. (2005). Serial position effects in recall of television commercials. *The Journal of general psychology, 132*(2), 151–164.

CHAPTER 9   I REMEMBER WHEN…OR DO I?

multiple-choice tests allow the same way that looking at shelves in a grocery store would help with the list) requires fewer cognitive resources than generating the information in its entirety (as is necessary with essay writing).

This difference, which is also related (somewhat) to the navigational preferences (folders vs. search) described in the "Cognitive Cartography" section, isn't just something that should be considered for students, however. Indeed, studies have shown that as we age it actually becomes more difficult (i.e., takes more cognitive resources) to recall information than to recognize it.[32]

This understanding informs Nielsen's well-known design heuristic[33] of supporting recognition over recall. To do this, we design the equivalent of the multiple choice test by providing a history of recently viewed/used items, comprehensive menus prioritized by frequency of use, and consistent locations/designs that leverage familiarity.

Recap

- Any information – for what something is, was, or did or how to do something – that we carry with us constitutes a memory.

- Memories last different amounts of time, ranging from a fraction of a second to a lifetime. As memories are deemed important, they become re-encoded in progressively longer-lasting memory stores, going from sensory to short-term/working to long-term.

---

[32] Rhodes, S., Greene, N. R., & Naveh-Benjamin, M. (2019). Age-related differences in recall and recognition: A meta-analysis. *Psychonomic Bulletin & Review*, *26*(5), 1529–1547.

[33] Nielsen, J., & Molich, R. (1990). Heuristic evaluation of user interfaces, Proc. ACM CHI'90 Conf. (Seattle, WA, 1-5 April), 249–256.

238

CHAPTER 9   I REMEMBER WHEN...OR DO I?

- Working memory is what we use to temporarily store information that we need in a given moment. This includes newly acquired information as well as information that we may be bringing out of long-term storage (i.e., is also "where" we remember things).

- Even though working memory has limitations on the number of items that can be simultaneously processed or thought about, there are some tricks we can use, like chunking information, to extend our ability to keep things in this short-term memory store.

- Understanding and leveraging how memory works helps make designs more effective and easy to use.

- Although long-term memory is practically limitless, selecting and storing information in these stores can be laborious and error-prone. A tried-and-true mnemonic tool to ease the storage of information is the method of loci, which exploits the evolutionary and biological underpinnings of long-term memory storage.

- Cognitive biases inform why we remember things poorly or inaccurately.

- Recognizing previously experienced information is much easier (i.e., recognition) than generating what was seen purely from our memories (i.e., recall).

Before You Go...

Most people's memories of Disneyland involve going on rides like Space Mountain, exploring the spooky (but not scary) fun of the Haunted Mansion, or perhaps eating some form of mouse-shaped treat.

239

CHAPTER 9   I REMEMBER WHEN...OR DO I?

And for the lucky few, there may also be memories of not just all of the above but the opportunity to interact with a costumed cast member, possibly even Bugs Bunny[34]!

"Um, wait a minute," may be your response right now, and if so, you're correct. Although Bugs Bunny, a trademarked character of the Warner Brothers company, would never be at the Walt Disney Company property in Anaheim, California, this memory is exactly what Dr. Elizabeth Loftus and her colleagues were able to implant in the minds of some research participants in a study they conducted on memory reconstruction. Because our act of remembering relies on our working memory system, anything additional that we encounter when remembering something from our past can become corrupted with new information. To this end, the research team asked student participants to read advertisements for Disneyland that required them to evoke their own memories, with the addition of a chance (and erroneous) encounter with a certain sassy carrot-eater who had no business being there. When asked to recall their own experiences at Disneyland (not the content of the ad), of those individuals who read the fake ad, almost one-fifth (16%) actually remembered hugging or meeting Bugs Bunny personally!

And if you think that such memory implantation could never happen to you (i.e., you'd be in the majority who wouldn't falsely remember hugging a cartoon character from a different studio), you should note that when the information to be implanted strains credibility less, even higher numbers of people remember doing something that never happened. For example, when research participants were presented with information that suggested they had "slimed" a teacher at one point in time when they were in school, almost two-thirds (65%) remembered doing so when

---

[34] Braun-LaTour, K. A., LaTour, M. S., Pickrell, J. E., Loftus, E. F., & Distinguished, S. U. I. A. (2004). How and when advertising can influence memory for consumer experience. *Journal of Advertising, 33*(4), 7–25.

asked about something inappropriate they had done when they were later tested.[35] So, yes, memory implantation is a legitimate phenomenon – just don't use this information for evil, because chances are good, some of your memories aren't what you think they are.

---

[35] Loftus, E. F. (2004). Memories of things unseen. *Current directions in psychological science, 13*(4), 145–147.

CHAPTER 10

Making Decisions: Why We Buy Lottery Tickets

RB: At least once a year, I have to go to Las Vegas for some conference or another, and consistently, everyone around me has loads of advice on how to gamble effectively.

JLR: Seriously?

RB: Yep. And just as consistently, I tell them I don't gamble, because I'm good at math.

JLR: Which is the opposite of what a lot of people think of themselves. Have you ever noticed that people (including some teachers) have no problem volunteering how they are "bad at" or "hate" math? This seems crazy to me, especially if you consider that for almost any other core academic subject, like reading, it would be considered not just in poor taste but sacrilegious?

RB: True! And in many ways, it's less about being bad or good at math and more about understanding why we make mistakes when numbers are involved. It's not that people are bad at math at all but rather that certain shortcuts we're wired to make interfere with our ability to think about certain types of math logically.

JLR: I'd give eight to one odds on that.

© Jerome L. Rekart and Rebecca Baker 2025
J. L. Rekart and R. Baker, *Designing for Human Intelligence in an Artificial Intelligence World*,
https://doi.org/10.1007/979-8-8688-1418-1_10

CHAPTER 10   MAKING DECISIONS: WHY WE BUY LOTTERY TICKETS

How Numbers Stymie (Some of) Us

Although many individuals claim that they aren't "numbers people" (which creates a great deal of consternation for both of us), if the universe has a language, it would be mathematics. Indeed, creatures ranging from honeybees to chickens to rats to chimpanzees have all been shown to be able to not only count but make decisions based on ordinality (i.e., they always choose the fourth arm of a maze because that is where they have learned the food to be, no matter how many arms there are).[1]

I CALCULATE, THEREFORE I AM

This notion of a universal language has been extended even further by some scientists. The most extreme logical conclusion of this premise has been made repeatedly by MIT physicist, Dr. Max Tegmark, who has put forth the Mathematical Universe Hypothesis (MUH), which suggests that the universe is actually a mathematical structure itself, whose physical representation is a byproduct.[2]

Thus, the ability to count is a part of being alive. We know the utility of math follows us throughout our life, from its beginnings of understanding "more" vs. "less" through our ability to calculate trajectories and speeds (e.g., as are necessary to drive or catch a baseball) and culminating in advanced systems like algebra, calculus, and statistics.

---

[1] Vallortigara, G., Chiandetti, C., Rugani, R., Sovrano, V. A., & Regolin, L. (2010). Animal cognition. *Wiley Interdisciplinary Reviews: Cognitive Science, 1*(6), 882–893.

[2] Tegmark, M. (2008). The mathematical universe. Foundations of physics, 38(2), 101–150.

CHAPTER 10   MAKING DECISIONS: WHY WE BUY LOTTERY TICKETS

And yet, most of us will make very specific mistakes when it comes to using numbers to guide us through our lives. It has been shown that the majority of people will be more likely to mistake the likelihood of an event (either over or under) when it is presented in terms of a percentage than when the same information is provided in terms of whole numbers, even though the two are mathematically the same.[3] Thus, people assume that an event that has a probability of 0.001 is rarer than one that will afflict 1 out of 1000 people. This effect is caused, in part, by the fact that we represent numbers very differently in our minds. The fact that animals and small children can engage in magnitude distinctions (less vs. more) and counting shows that higher-level cognition is not a requirement of some mathematical abstractions. This initial "gut-level" (although of course it is taking place in the brain) process of thinking about numbers bears out in many different ways.

For example, smart marketers know that $100 represents a threshold that changes how you advertise additional fees. Namely, any time you have to represent an additional fee that is less than $100, you should do so in terms of dollars (an amount), but when you are doing so on top of a principal amount that is greater than $100, you should use a percentage.

The reason this works is because humans have a "feeling" for how large numbers are, which is referred to as the analog (or analogue in the United Kingdom) magnitude system.[4] Our analog magnitude system is a mental number line that we use to determine how large something *feels* rather than how big it actually is. Thus, for anything less than $100,

---

[3] Kahneman D, Frederick S. 2002. Representativeness revisited: attribute substitution in intuitive judgment. In Heuristics and Biases: The Psychology of Intuitive Judgment, ed. T Gilovich, D Griffin, D Kahneman, pp. 49–81. New York: Cambridge Univ. Press.

[4] Weathers, D., Swain, S. D., & Carlson, J. P. (2012). Why consumers respond differently to absolute versus percentage descriptions of quantities. *Marketing Letters*, *23*, 943–957.

CHAPTER 10   MAKING DECISIONS: WHY WE BUY LOTTERY TICKETS

the same amount will look smaller if it is presented as an absolute dollar amount rather than a percentage, for example:

$85 + $12.50 shipping and handling

will result in increased sales relative to.

$85 + 14.75% shipping and handling

However, the opposite is true for anything other than $100, such that one would be wise to advertise:

$275 + 14.75%

rather than (even though both are exactly the same dollar amount)

$275 + $40.56

Even when participants are presented with both values and asked which one feels smaller, the majority of individuals will identify the dollar amount for quantities less than $100 and the percentage for those greater than it.[3]

This gut-level means of handling numbers becomes even more pronounced when we are required to make predictions. For example, in a naturalistic observation, it was found that when numbers "hit" (i.e., were part of the winning series of a lottery) the previous week, they are *less* likely to be played the following week; however, when they "hit" for multiple weeks in a row (are among winning tickets for two or more weeks), the likelihood that they get played in later weeks actually *increases*.[5] Thus, let's say in a game where 6 numbers are randomly selected from a pool of 49

---

[5] Galbo-Jørgensen, C. B., Suetens, S., & Tyran, J. R. (2016). Predicting Lotto Numbers A natural experiment on the gamblers fallacy and the hot hand fallacy. *Journal of the European Economic Association, 14*(3), 584–607.

CHAPTER 10   MAKING DECISIONS: WHY WE BUY LOTTERY TICKETS

(with no duplication), in week 1, a winning ticket that pays in full looks like this:

1, 49, 5, 30, 3, 44

The first finding means that of the numbers played the next week, the other 43 numbers that didn't win in week 1 (above) will appear on more tickets than they normally would (put differently, that numbers 1, 49, 5, 30, 3, and 44 will be played less often the following week). This is called the "gambler's fallacy."

However, let's say that in week 2 (following the winning ticket above), the following numbers would be found on a winning ticket:

17, 45, 13, 11, 6, 3

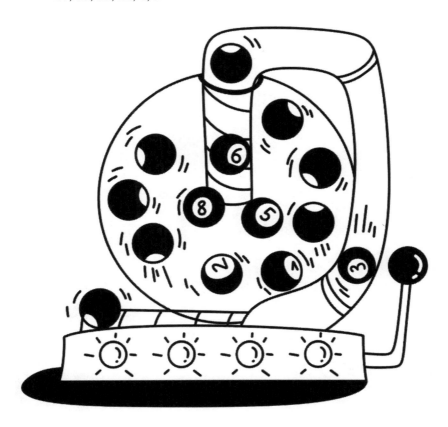

CHAPTER 10   MAKING DECISIONS: WHY WE BUY LOTTERY TICKETS

Now, because the number 3 "hit" in two consecutive weeks, it will be *more* likely to appear on tickets played the following week (week 3) than would normally be the case (and the other numbers, 17, 45, 13, 11, and 6, will again be less likely). This latter phenomenon is referred to as the "hot-hand bias."

These seemingly contradictory behaviors speak to the fact that many humans think that underlying numeric or probabilistic patterns should be observable in all samples of that behavior. Put differently, if some overarching phenomenon is truly random, then the belief dictates that any sample of the behavior should also *look* random. This fallacy, which was coined "belief" in the "Law of Small Numbers,"[6] is at play in the lottery examples just described. In the first occurrence, people mistakenly believe that somehow the numbers from the first week will make their appearance the following week less likely, despite the complete independence of bouncing ping pong balls making their way into the exit tube as "winners." The second is that there is an innate sense that somehow randomness is violated if a number comes up repeatedly in subsequent weeks, suggesting that somehow the small sample is providing enough data to prove that there is something special about the winning number(s).[7]

---

[6] Tversky, A., & Kahneman, D. (1971). Belief in the law of small numbers. *Psychological bulletin, 76*(2), 105–110.

[7] Benjamin, D. J. (2019). Errors in probabilistic reasoning and judgment biases. Handbook of Behavioral Economics: Applications and Foundations 1, 2, 69–186.

CHAPTER 10   MAKING DECISIONS: WHY WE BUY LOTTERY TICKETS

I'M FEELING LUCKY!

The same is sometimes (equally erroneously) applied to people. My father-in-law likes to play the lottery and, in fact, has had some decent winnings over the decades that he has indulged his dollar or so a week habit. Does this mean that he is inherently "lucky" – he (and I) agree that "no," there is no system or special divine intervention that has bestowed him with winnings above what one may consider to be outside of chance. However, there are some individuals who believe that one can have a "hot hand" with games of chance. This is another example of the law of small numbers at work whereby people latch on to what seem to be deviations from randomness as proof that there is something greater (or more beneficial) happening. Because there is no special formula to cheat chance, I do not gamble; however, I cannot argue with father-in-law's mantra, which is that "you can't win if you don't play." I just believe, which is equally true, that you can't lose either.

Indeed, the latter situation has been shown to apply to a number of different situations. For example, it has been shown that when a family consists of three (or more) children of the same sex (i.e., three daughters or three sons), a majority of individuals will state that having a fourth of the same sex (e.g., another daughter, if there are already three) is the most likely outcome. Contrary to this, if you ask people to consider a family with an equal number of boys and girls (e.g., two daughters and two sons) and ask them to predict the likely sex of the next child for that family, over 90% of responses will indicate that the chance of either a male or a female is the same.[8] Though this latter response suggests that there is a foundational

---

[8] McClelland, G. H., & Hackenberg, B. H. (1978). Subjective probabilities for sex of next child: US college students and Philippine villagers. *Journal of Population*, 132–147.

CHAPTER 10   MAKING DECISIONS: WHY WE BUY LOTTERY TICKETS

understanding of the equality of odds (all things being, well, equal), it should be noted that some researchers have found that many individuals believe that if systems are random, then they will "correct themselves" to maintain this randomness.[4] For this to be true, it would mean that somehow the necessary male germ cell (i.e., the sperm is responsible for whether a child is male or female because the egg has two X chromosomes) would be "allowed" to fertilize the egg to reset the proportion of boys and girls at 50:50! Again, this belief in the law of small numbers makes itself felt!

To illustrate what this means, let's assume that there is a hospital (de Laplace Memorial) that normally assists with the delivery of 100 children a month, and it is now halfway through a month that has seen the birth of 35 girls and 15 boys (for a total of 50). When asked what the likely distribution of boys and girls will be for the predicted 50 additional births that month, studies have shown that a majority of respondents will be more likely to predict that 35 boys and 15 girls will be born (thus resulting in a total of 50 boys and 50 girls for the month). Again, people will give this response even though the correct answer (again, all things being equal) is that the most likely outcome will be 25 boys and 25 girls (meaning that for that month, there would be 60 girls and 40 boys born). It is only over the long run that probabilities are accurately observed.

The types of probabilities generated by people in these settings are called "subjective probabilities,"[9] so named because they rely on calculations based on consistency with underlying assumptions about situations, such as what randomness "looks" or "feels" like, rather than actual mathematical calculation, which are called "objective probabilities" (or just, well, probabilities).

Phenomena such as the gambler's fallacy and hot-hand bias illustrate how much "feeling" enters into our unconscious processing of quantitative information. It *feels* like we shouldn't play the same number two weeks in a

---

[9] Anscombe, F. J., & Aumann, R. J. (1963). A definition of subjective probability. *Annals of mathematical statistics, 34*(1), 199–205.

CHAPTER 10   MAKING DECISIONS: WHY WE BUY LOTTERY TICKETS

row as that would violate randomness, UNLESS the same number hits two (or more) weeks in a row, at which point it "feels" like there is something special about that number that causes it to exist outside randomness.

Feelings such as these also cause us to misjudge the likelihood of other events.

In a now famous set of experiments,[10] research participants were presented with brief descriptions of individuals (in the original study they were "Bill" and "Linda"). With only those descriptions (an example is given below), the participants were then asked to rank the likelihood that Bill or Linda would engage in or be defined by a given set of attributes (jobs and hobbies).

For example, we could present the following description about a different individual, Danny:

> Danny is 37 years old. He is intelligent, but not generally creative. In school, he was strong in mathematics but weak in the humanities.

And then we could ask you to determine how representative the following attributes are by ranking them from most to least. As you read through the descriptions, ask yourself, which do you see Danny doing or being and which do you not?

- Danny is a physician who plays poker.

- Danny is an accountant.

- Danny is an architect.

- Danny surfs for a hobby.

---

[10] Tversky, A., & Kahneman, D. (1983). Extensional versus intuitive reasoning: The conjunction fallacy in probability judgment. *Psychological review*, *90*(4), 293–315.

251

CHAPTER 10 MAKING DECISIONS: WHY WE BUY LOTTERY TICKETS

- Danny plays in a Grateful Dead cover band.

- Danny is a reporter.

- Danny climbs mountains for a hobby.

- Danny is an accountant who plays in a Grateful Dead cover band.

As you read through these descriptions, you may have felt like they were leading you in a particular direction. That is to say, some of the descriptions play both into logical conclusions that someone who would be good at math might be drawn to accounting and stereotypes that accountants may not be "creative" (again, this is a stereotype; although there are some financial experts in white collar prisons for just that conjunction: creative accounting).

The descriptions that seem to follow from those conclusions and/or stereotypes were actually the only ones that the researchers were interested in seeing participants' responses to. This means that even though eight different statements were provided, the researchers were actually only interested in three of them (in bold below). The researchers used the same setup for various hypothetical individuals; though they varied the names and attributes, they were always only interested in three of the descriptions: a job or hobby that was very likely, one that was unlikely, and the combination of the two. Thus, the three that would have been of interest are

- Danny is a physician who plays poker.

- **Danny is an accountant** (**A**).

- Danny is an architect.

- Danny surfs for a hobby.

- **Danny plays in a Grateful Dead cover band** (**GD**).

CHAPTER 10 MAKING DECISIONS: WHY WE BUY LOTTERY TICKETS

- Danny is a reporter.

- Danny climbs mountains for a hobby.

- **Danny is an accountant who plays in a Grateful Dead cover band (A+GD).**

In this case, **A** (Danny is an accountant) is the most likely, **GD** (Danny plays in a Grateful Dead cover band) would be the unrepresentative statement, and **A+GD** (Danny is an accountant who plays in a Grateful Dead cover band) is the combination of the two.

Remember that the task was to rate the relative likelihood that a given description was representative of Danny. Thus, what the researchers were interested in was how likely A was relative to GD relative to A+GD. What they found was that for over 80% of participants – regardless of the particular descriptions – that the following order was always found:

Likely > Combination > Unlikely

Is that what you selected for Danny? Meaning did you feel like him being an accountant (**A**) had the best chance of describing him accurately and him playing in a Grateful Dead cover band (**GD**) was the least likely? Put differently, did you have:

A > A+GD > GD?

If so, then you, like the overwhelming majority of others, went with a gut feeling (i.e., a subjective probability) rather than the actual statistical likelihood and committed what is called the "conjunctive fallacy." Therefore, with the information you were given, an accurate answer would be that the number of accountants is likelier than Danny playing in a Grateful Dead cover band, which is more likely than him being an accountant who plays in a cover band:

A > GD > A+GD

253

CHAPTER 10 MAKING DECISIONS: WHY WE BUY LOTTERY TICKETS

And, because no actual numbers were provided and seeing as we were using stereotypes to make our assignments, technically, any of the following could also be possible (though, admittedly improbable):

$$A \geq GD > A+GD$$

or

$$GD > A > A+GD$$

or

$$GD \geq A > A+GD$$

Indeed, the only order that isn't EVER possible is the one that is most often provided (as shown above):

$$A > A+GD > GD$$

So, why is Danny more likely to be in a Grateful Dead cover band and NOT be an accountant than be a CPA (or CMA) who also gets lost in 60-minute renditions of "Fire on the Mountain" on the weekends? Because the chances of a conjunction, which is the likelihood of both events occurring (i.e., in this case A and GD), must always be less than the chance of either of the events happening on their own. Think about the fact that there are many more people who play in Grateful Dead cover bands who are not accountants (GD) than there are people who are accountants and who play in Grateful Dead cover bands (A+GD; see figure).

254

CHAPTER 10   MAKING DECISIONS: WHY WE BUY LOTTERY TICKETS

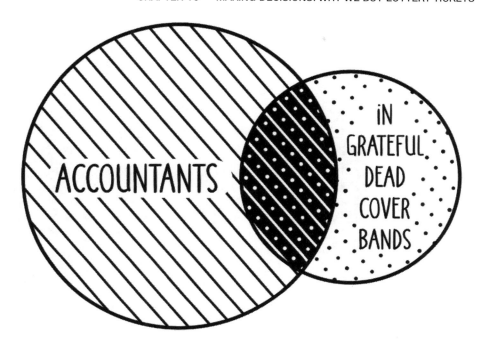

People get drawn into making conjunctive fallacies, like the other biases we discussed at the opening of this chapter, because of feelings. We can find ourselves getting anchored to one particular element (Danny must be an accountant) and then fail to take into account other elements (like logic) and conclude that the next likeliest possibility is that he is an accountant who does something else.

Data and Design: Don't Go with Your Gut

One of the elements we keep coming back to is "know your users." The conjunctive fallacy shows us that we need to be aware of biases that might creep into our designs or the strategies we employ to meet them when we fail to consider actual data or go off of our feelings rather than facts. This is one downfall of using personas. Though our use of personas can help us to really think about the problems and needs of various groups of

CHAPTER 10   MAKING DECISIONS: WHY WE BUY LOTTERY TICKETS

users, we need to be careful that we don't get caught up in who we think these personas represent (remember: a persona is not a real person) and examine the actual numbers of (real) people who would be impacted by a particular feature or navigational tweak.

For example, let's say that you need to revise your current shopping basket design to make it easier to change individual items within a category (e.g., replace one brand of detergent with another that is of equal or lesser price) by reducing the number of steps, which currently requires a user to:

1. Remove the out-of-stock item from the basket

2. Return to the shopping screen

3. Search for a comparable brand

4. Place that item in the basket as a replacement and

5. Return to the checkout screen

We'll call the new design we are working on "Replace" (R). You know from user feedback that this feature will help many people, but you decide to go into your personas to see if there is anything else you should be considering. While looking through your files, you note that in the "About" section for "Chloe" (one of your user personas), it mentions that she is very "eco-conscious" and makes purchasing decisions based on this. We'll call this attribute of being concerned about the environment "ECO." Now, what is clear here is that your primary design task (i.e., will help the greatest number of users) is "R." What you need to ensure, however, is that you don't erroneously presume that

$$R > R + ECO > ECO$$

where R+ECO means considering a redesign of a basket that weighs costs (the primary design requirement) and ecological consciousness (R+ECO). Why could this be a problem?

CHAPTER 10   MAKING DECISIONS: WHY WE BUY LOTTERY TICKETS

In this situation, a conjunctive fallacy would cause you to overinflate the importance and size of the user base for whom R+ECO might be preferred. In this case, considering designs that would meet the requirements of R+ECO (provide replacements that are the same price AND ecologically responsible) might skyrocket costs and greatly increase timelines as you try to consider ways to provide replacements in the basket that are the same price AND are purportedly as "green." The resources required to do this may not be warranted given that the number of actual users who would want that latter may not in actuality be that numerous (i.e., R+ECO must be the smallest group). Indeed, it may be that when "ECO" is a primary driver or motivator for someone, they are willing to pay more to maintain that level of eco-consciousness, even when their preferred product brand isn't available.

TWO FOR ONE, OR NOT

Conjunctive fallacy is an incredibly common problem when designing new features. Remember the story about trying to take attendance for our teachers? In the midst of solving this problem, a product manager consistently brought up that teachers also needed to be able to identify students that require interventions and why not enable both of those things on the screen. While they were correct (teachers did need to take attendance and identify students with interventions), they were also conflating the two priorities. Attendance-taking was a required, time-bound task, tied directly to the school's funding. Identifying intervention needs was also required, but not time-bound or tied to funding. As such, it made sense to pursue these two pieces as separate features, rather than bundling them into a single effort (and screen).

CHAPTER 10   MAKING DECISIONS: WHY WE BUY LOTTERY TICKETS

Is the Juice Worth the Squeeze?

In the world of business, determining whether or not something is worth doing (i.e., will the time and money needed to create a thing result in sales that are greater than the initial and ongoing investment) may seem straightforward. Unlike the philosophical considerations of utilitarian ethics wherein the question of morality is tied to the utility of the action vs. its investment,[11] cost-benefit analysis in business decisions seems straightforward. Or is it? Like so many other things we've discussed, it's complicated. And, like the other things we've discussed in this book, it's important to understand when designing products – both for how you advocate (or deprecate) design decisions for investment and for how you present investment options and opportunities for your user.

---

[11] Kelman, S. (2013). Cost-benefit analysis: an ethical critique. In *Readings in risk* (pp. 129-138). RFF Press.

CHAPTER 10   MAKING DECISIONS: WHY WE BUY LOTTERY TICKETS

The first layer of complexity comes with identifying costs. Generally speaking, costs fall into four categories:

- **Direct**: How much money and time it will take to implement the idea.

- **Indirect**: What it takes to keep the lights on, like renewing software license costs, utility payments, etc.

- **Intangible**: This is the bucket of "squishy measures." Intangible costs include things like lost advancement on project A because your only developer is working on project B (the new thing), lost sales from customers who didn't like a process change, and so on.

- **Opportunity**: To pursue any opportunity, you have to NOT pursue other opportunities. What is the cost of abandoning those opportunities in favor of the one you chose?

Now that you've laid out these costs, the problem becomes obvious – not all of these things are the same. We cannot easily quantify a lost opportunity or potential customer dissatisfaction. And, when doing calculations, it's important to have a common scale – a way to connect and normalize all of the costs so that they can be combined and compared to the benefits (spoiler alert: benefit measures have to use the same scales as the costs). This is the tricky part where a lot of companies just throw up their hands and say "Screw it, we're going with direct and indirect and use our gut for the rest." And based on the previous sections, how good is our gut? Exactly. So what can we do? Well, as you may have guessed based on the previous nine chapters, our answer here is research. Taking the time to do market research on things like willingness to pay, current market trends, and surrogate indicators can provide a level of confidence in these measures, letting you express them in the same terms (monetary) as the more easily measured direct and indirect costs.

259

CHAPTER 10   MAKING DECISIONS: WHY WE BUY LOTTERY TICKETS

Once you have identified (and normalized) your costs, you can quantify your benefits. These also fall into four categories:

- **Direct**: Revenue from new or increased sales

- **Indirect**: Better street cred (a.k.a. reputation) for your company, more efficient processes, and similar

- **Intangible:** Being seen as a thought leader/innovator, improved employee happiness, etc.

- **Competitive:** First to market, best in market

As you can see, once again, these are not easily quantified using the same scale. Market research and careful data analytics can help – specifically techniques like inferential statistics. Having gathered both your costs and benefits, you can now compare and use the results to help inform your decision whether or not to go forward. Note we said "help" inform your decision. While cost–benefit analysis is a great way to ensure you are letting the data drive your decisions, it should never be used as the only thing that drives the process. Cost–benefit analysis has numerous pitfalls such as difficulty in predicting all the potential variables and an inability to quantify the ethical or moral imperative behind any particular decision.

These considerations are important from a very practical "should we build a thing" perspective, but what about your user? Users are constantly making these types of decisions about your products, but without the benefit of extensive market research or a data team to apply inferential statistics. Instead, they are relying on their gut, something that varies with the individual based on their current context, past experiences, and even age.[12] To guide the user in their decisions, we need to understand a bit more about how humans think about certainty vs. uncertainty.

---

[12] Löckenhoff, C. E. (2011). Age, time, and decision making: From processing speed to global time horizons. *Annals of the New York Academy of Sciences, 1235*(1), 44–56.

260

CHAPTER 10   MAKING DECISIONS: WHY WE BUY LOTTERY TICKETS

Decisions Under Certainty vs. Uncertainty

Hopefully we have made it crystal clear that humans are not rational decision-makers even under the best of circumstances. Though this is something that we now take for granted, economists and psychologists once considered the average individual to be a rational decision-maker, meaning that they were believed to consciously and carefully weigh all possible outcomes before arriving at decisions. This philosophy was formalized in the 1940s by von Neumann and Morgenstern in their "expected utility theory"[13] whereby they posited that all things being equal, the typical individual makes decisions based on what will provide them with the greatest value or "utility." They presupposed in their models that humans not only seek to maximize the result of their decision (i.e., a rational decision) but do so without other factors at play. Subsequent to this work, the psychologists Amos Tversky and Daniel Kahneman demonstrated how irrational decisions are made by participants in laboratory settings: findings which were buttressed by evidence of many poor choices made in the real world. In fact, this work so revolutionized the way we now think about decision-making that Dr. Kahneman was awarded the Nobel Prize in Economics in 2002.[14]

In place of expected utility and rational decision-making, Kahneman and colleagues put forward a model of subjective utility they dubbed "prospect theory,"[15] which set out to show how various factors, not the least of which being risk, affected the ways in which people see various choice

---

[13] Frisch, D., & Clemen, R. T. (1994). Beyond expected utility: rethinking behavioral decision research. *Psychological bulletin, 116*(1), 46–54.

[14] Dr. Tversky passed away in 1996 and thus wasn't a corecipient as the Nobel Committee does not award the prize posthumously.

[15] Kahneman, D., & Tversky, A. (2013). Prospect theory: An analysis of decision under risk. In *Handbook of the fundamentals of financial decision making: Part I* (pp. 99–127).

261

CHAPTER 10 MAKING DECISIONS: WHY WE BUY LOTTERY TICKETS

points and decide between them. Think about your local grocery store. Up and down the aisles are all sorts of products packaged in such a way to not only grab your attention but persuade you to buy them rather than their competitors. Given the health consciousness of today's consumers (at least aspirationally), most products are marketed in such a way that the salubrious elements are highlighted and the negative or unhealthy elements are downplayed. For example, ground meat (beef, chicken, turkey) is referred to by how lean it is (conveying healthiness) rather than how much fat it has (conveying unhealthiness). If expected utility theory were the correct model to describe human behavior, then people should be equally likely to choose either of the following two products:

A: 75% fat-free ground beef

B: 25% fat ground beef

However, in numerous studies (and in the real world), people not only overwhelmingly indicate that they would be more likely to buy product "A" (which DOES contain 25% fat), they report that product B not only seems less healthy but that it tastes greasier and is overall less desirable.[16] The fact participants and consumers feel this way despite knowing that the two things are in essence the exact same speaks to the power of how one frames the parameters of a decision.[17] By framing a product, like ground beef, in terms of "fat," there is an activation of mental networks associated with oleaginous unhealthiness rather than fitness and health. Indeed, the

---

[16] Flusberg, S. J., Holmes, K. J., Thibodeau, P. H., Nabi, R. L., & Matlock, T. (2024). The psychology of framing: How everyday language shapes the way we think, feel, and act. *Psychological Science in the Public Interest*, *25*(3), 105–161.

[17] Kahneman, D. (2013). A perspective on judgment and choice: Mapping bounded rationality. *Progress in Psychological Science around the World. Volume 1 Neural, Cognitive and Developmental Issues.*, 1–47.

CHAPTER 10   MAKING DECISIONS: WHY WE BUY LOTTERY TICKETS

higher the stakes that are framed, the more people can be nudged toward making one decision over another.

In their 1981 paper, Tversky and Kahneman presented all of the research participants with a hypothetical scenario whereby an outbreak of a disease would be expected to kill 600 people. They then presented half of the participants (Group 1) with two potential programs and asked them to choose the one that they felt most comfortable with. The programs presented to the first group were

A1: 200 people will definitely be saved.

B1: One-third probability that 600 will be saved and a two-third probability that nobody will be saved.

The other half of the participants (Group 2) were instead given the choice between these two programs:

A2: 400 people will die.

B2: One-third probability that nobody will die and a two-third probability that 600 people will die.

Now, (hopefully) you noted that in terms of the number of folks who would survive, A1=A2 and B1=B2.[18] You probably also noticed that for the first group (A1 and B1), both programs are framed in terms of saving people, but for the second group (A2 and B2), they are framed in terms of the number who would die. Thus, the only difference between the two is in terms of how the programs are framed.

Given these differences it should come as no surprise that the majority of respondents in the first group chose Program A1 over B1 by a difference

---

[18] And, if you've had probability theory, then you may have also noted that in terms of the chances of a given outcome, in fact A1=A2=B1=B2, whereby the outcome (i.e., likelihood x number) for the four is as follows: A1/A2: 100% chance of being saved x 200 people = 200; B1/B2: 33% chance of being saved x 600 people = 200.

CHAPTER 10   MAKING DECISIONS: WHY WE BUY LOTTERY TICKETS

of 72% vs. 28%; however, in the second group, the majority of participants chose Program B2 over A2 (78% for B2, 22% for A2).

These data, and the many replications that have followed, reinforce that the manner in which information is presented matters. They also show that people make decisions differently when presented with a situation that is framed in terms of a negative outcome (400 people will die) vs. a positive one (200 people will survive). In the former situation, a gamble is preferred by most people (i.e., go with the risk) because the thought of assured death is evoked, whereas in the latter situation, the idea that some folks will definitely survive, a positive outcome, is the preferred one – even though it results in the exact same negative result for 400 people. These results have been framed as showing that when it comes to gains, people are primarily risk averse, meaning that they would rather take the sure thing. However, when it comes to losses (like the ultimate one), the majority of individuals will opt to try their chances.

If you were curious why a psychologist studying decision-making was awarded the Nobel Prize in Economics, it was in part because his studies (and subjective utility curves) were able to explain why people sold their shares in companies soon after their stocks started to turn profits (i.e., take the sure thing) but would hold on to plummeting shares for way too long when they were losing money, thus compounding their losses (i.e., take one's chances that the market would swing back up again).

Neuroscientists have been able to use modern imaging techniques to view what happens in the brains of individuals when they're asked to make decisions such as these. Not surprisingly, these studies have shown a pronounced activation of emotional centers within the brain, which explains the impact that emotionality has on decision-making.[19]

---

[19] De Martino, B., Kumaran, D., Seymour, B., & Dolan, R. J. (2006). Frames, biases, and rational decision-making in the human brain. *Science, 313*(5787), 684–687.

CHAPTER 10   MAKING DECISIONS: WHY WE BUY LOTTERY TICKETS

And provides the last nail in the coffin (or does it fully seal the lid?) for the misguided notion that humans make rational decisions in all circumstances.

Scarcity and Abundance: Impacts on Decision-Making

One advantage of ecommerce for sellers is the ease with which they can create an *appearance* of scarcity. Indeed, when one's goods are not on full display, it is much easier to claim that stock is running low (though to do so when it isn't true is certainly unethical). This is critical because early research on product placement and consumer behavior showed that when demand is high and supplies are low (i.e., scarcity conditions of availability), consumers become stressed (and we just covered what happens to decision-making when emotion gets involved). This duress is advantageous for stores, because it increases cognitive load, which in turn depletes mental resources that customers might otherwise use to carefully weigh purchasing decisions.[20] (Remember our dark pattern of false urgency?)

Other forms of scarcity also have pronounced impacts on consumer behavior. For example, temporal constraints pressure someone to make a decision within a time frame that may be faster than they would normally prefer (i.e., scarcity of time). For example, old commercials on television would suggest that if you "act now," you could get a deal that otherwise wouldn't be available to you. In the modern era of ecommerce, the same idea takes the form of "lightning deals," "flash sales," and "limited time offers," all of which can positively impact consumer behavior in the short

---

[20] Godinho, S., Prada, M., & Garrido, M. (2016). Under pressure: an integrative perspective of time pressure impact on consumer decision-making. *Journal of International Consumer Marketing, 28*(4), 251–273.

265

CHAPTER 10 MAKING DECISIONS: WHY WE BUY LOTTERY TICKETS

term (i.e., result in greater sales), though they may not result in satisfied customers in the long term who end up feeling regret over their decision.[21] As with perceptions of product shortages ("only three items left in stock!"), feelings of time closing in (e.g., "ACT NOW!") increase cognitive load, which reduces rational processes, such as those that would be used to carefully weigh advantages and disadvantages of potential options. Indeed, in a web setting, this increase in cognitive load due to perceived scarcity also reduces the effectiveness of online product reviews (from past customers), which reflects the idea that under duress, individuals are more likely to participate in risky behavior.[22]

Not only can web-based marketplaces introduce perceptions of scarcity much easier than can be done in brick-and-mortar settings, they can also conversely provide seemingly unlimited options from which to choose (i.e., product abundance). With no physical limitations like shelf space or traffic areas to consider, e-retailers need only dedicate space on a virtual carousel to a myriad and sundry number of potential variations on a theme. Indeed, a recent search for winter gloves turned up an enormous number of options on the leading US-based (but South American-named) marketplace, with over six pages of results with each page listing between 60 and 70 different products.

But given everything that we've discussed about cognitive load, is the "everything and the kitchen sink" approach to ecommerce the best way to move merchandise? Meta-analyses suggest that for the vast majority of sellers (i.e., those who do not have massive infrastructure and networks

---

[21] Barton, B., Zlatevska, N., & Oppewal, H. (2022). Scarcity tactics in marketing: A meta-analysis of product scarcity effects on consumer purchase intentions. *Journal of Retailing, 98*(4), 741–758.

[22] Wrabel, A., Kupfer, A., & Zimmermann, S. (2022). Being informed or getting the product? How the coexistence of scarcity cues and online consumer reviews affects online purchase decisions. *Business & Information Systems Engineering, 64*(5), 575–592.

CHAPTER 10   MAKING DECISIONS: WHY WE BUY LOTTERY TICKETS

of third-party vendors), there *are* limits to the number of choices that the average consumer can absorb before becoming overwhelmed. The term for these feelings is "choice overload," and a recent meta-analysis (i.e., a rigorous examination and study of other studies that have been conducted to find consistent trends) showed that indeed having too many items can be counterproductive to decision-making and result in what is called "deferral," which means closing a browser before completing a sale.[23] There are four parameters in particular that must be considered when contemplating whether there are too many items being presented to users. As each of the following parameters changes, so too does choice overload:

- **Decision Difficulty**: Are there too many options to weigh? For winter gloves, there aren't so many, but for something like a car or mobile phone, there are likely many more parameters to consider, which increases the difficulty of the decision and choice overload.

- **Variability Between Choices**: Is there a standout product (referred to as a "dominant option") that acts as an anchor against which all others can be compared, or is the variability between the options barely noticeable, if at all? When there is little to differentiate one option from another, choice overload increases.

- **Consumer Preference**: Does the user know what she wants from a given product? Her uncertainty will become exacerbated as the number of options increases, which in turn is likely to result in choice overload.

---

[23] Chernev, A., Böckenholt, U., & Goodman, J. (2015). Choice overload: A conceptual review and meta-analysis. *Journal of Consumer Psychology, 25*(2), 333–358.

267

CHAPTER 10   MAKING DECISIONS: WHY WE BUY LOTTERY TICKETS

- **Decision Pressure**: When individuals are looking to minimize cognitive effort and just make a decision, the likelihood of choice overload will be increased concomitant with the number of available choices.

Foraging for Information

Remember how in Chapter 9 we mentioned that working memory might just be consciousness? You may have also noticed that even if it isn't responsible for our experience of being, it definitely influences our ability to make decisions. Many of the parameters we've described as impacting decision-making all come back to whether or not (and to what extent) cognitive resources are depleted. Indeed, the big (huge) takeaway from this chapter should be that design must consider not only how beautiful something looks and flows and feels but also the extent to which it accounts for cognitive load and, where necessary, exploits it.

Though we may not think of it as such, web searching is essentially foraging. Even if we are passing the time (or avoiding work) by going down some deep, dark Internet rabbit hole to find out as much as we can about polydactyl cats (nope, not related to pterodactyls), we are still – in terms of our brain – foraging. This distinction is important because when we engage in foraging – be it for information (e.g., about many-toed felines) or food – we evolved to strive to expend fewer resources than we collect. Thus, just as it doesn't make sense to exert hundreds of calories to swim to a patch of land that might contain three carrots (<100 calories total), we are also wired to not spend a great deal of cognitive energy searching through information we may or may not use.

Along these lines, studies have shown a primacy effect as it relates to embedded links (i.e., internal pages) on commercial websites. This means that no matter what the content, there is a bias to look through the topmost links (when vertically arrayed; leftmost when horizontally) that declines with the number of links. Indeed, in a rigorous real-life study (i.e., an

268

CHAPTER 10   MAKING DECISIONS: WHY WE BUY LOTTERY TICKETS

actual commercial site), researchers examined every possible combination of embedded links for a popular restaurant and found that whatever link was listed first was the one that received the most clicks.[24] Keep in mind that this means that even though "menu" may have been listed fourth (or last) in some of the versions, the overall trend was still for folks to click on whatever link was listed at the top of the list (e.g., "location"). This should not be a surprise as marketers have been clamoring for ideal shelf locations for decades (maybe centuries) and editors know to place stories that sell top-of-fold for print versions of newspapers (younger readers may want to Google what that is).

Influencing Perception and Persuasion

Over and over again, we are exposed to marketing and sales techniques that seem like they shouldn't work (especially because we are ALL aware of them) and, yet, they DO work. For example, campaigns making promises like the following:

> "And it can be all yours for only $19.99."

> "And if you act now, we'll send you not just one, but TWO dingle foppers!"

These work because the first taps into our reluctance to actually consider what numbers mean (see above). Thus, the last "9" of a decimal as a fraction and the last "0" as a whole (thus, "19.99" is perceived as being MUCH less than "20.00" than the actual 1/100th differential).[25]

---

[24] Murphy, J., Hofacker, C., & Mizerski, R. (2006). Primacy and recency effects on clicking behavior. *Journal of computer-mediated communication, 11*(2), 522–535.

[25] Liang, J., & Kanetkar, V. (2006). Price endings: magic and math. *Journal of Product & Brand Management, 15*(6), 377–385.

CHAPTER 10   MAKING DECISIONS: WHY WE BUY LOTTERY TICKETS

This phenomenon is also called the "left-digit bias" and is a result of focusing on the first set of numbers as a whole number "anchor."[26] And for the second example, it turns out that we just can't get past the notion that we're getting something for nothing,[27] even though if pressed most would concede that they understand that the cost of the "bonus" they are receiving is already factored into the price that is presented.

There are four standard persuasive techniques that are used to try and facilitate some kind of commitment – be it time or money – on the part of a consumer.

Foot in the Door

Have you ever felt like you ended up buying more than you anticipated? When it occurs deliberately, this is referred to as the "foot-in-the-door technique," and it involves making a small request that is then followed by a larger request (which is what the requestor actually wants you to do). This is believed to work because once someone initially agrees to the first request (the small one, or the "foot in the door"), a sense of commitment is created that makes them much more likely to consent to other, even larger, requests[28] that may then follow.

---

[26] Sokolova, T., Seenivasan, S., & Thomas, M. (2020). The left-digit bias: when and why are consumers penny wise and pound foolish?. *Journal of Marketing Research, 57*(4), 771–788.

[27] Lee, S., Moon, S. I., & Feeley, T. H. (2019). The "that's-not-all" compliance-gaining technique: when does it work?. *Social Influence, 14*(2), 25–39.

[28] Burger, J. M. (1999). The foot-in-the-door compliance procedure: A multiple-process analysis and review. *Personality and social psychology review, 3*(4), 303–325.

CHAPTER 10   MAKING DECISIONS: WHY WE BUY LOTTERY TICKETS

An example of this was demonstrated by Freedman and Fraser who called housewives and initially asked them if they would be willing to take a brief survey. Of those who said "yes," 52% then agreed – when asked a few days later – to participate in a much longer survey that would actually take place in their homes (obviously a MUCH larger commitment of time and planning). In contrast, when individuals weren't contacted beforehand with the quick phone survey, fewer than a quarter (22.2%) actually agreed to the in-home research.[29]

Door in the Face

Curiously, one can also reverse this strategy and, if executed properly, get the same result! This strategy, which is called the "door in the face technique," begins with a quite large (bordering on unreasonable) request with the expectation that it will be denied. After this failure to get what was asked for, the requestor then moves in to ask what they actually want, which is a much less involved (and more reasonable) behavior (e.g., like volunteering time) or a lower price point. It isn't actually the first request on its own that is persuasive but the contrast between the first unreasonable request and the second more realistic one. This is because that discrepancy makes the requestee (i.e., the potential customer or volunteer) feel bad that they can't help.[30]

---

[29] Joule, R. V., Girandola, F., & Bernard, F. (2007). How can people be induced to willingly change their behavior? The path from persuasive communication to binding communication. *Social and Personality Psychology Compass, 1*(1), 493–505.

[30] Cialdini, R. B., Vincent, J. E., Lewis, S. K., Catalan, J., Wheeler, D., & Darby, B. L. (1975). Reciprocal concessions procedure for inducing compliance: The door-in-the-face technique. *Journal of personality and Social Psychology, 31*(2), 206.

CHAPTER 10   MAKING DECISIONS: WHY WE BUY LOTTERY TICKETS

DIALING FOR DOLLARS

When I was a college freshman, I had a campus job that involved calling alumni to ask for donations for the university. I never understood at the time why the floor managers were so adamant that we "stick to the ladder," which required that we start at the "top" with what seemed like an absolutely crazy amount of money ($10,000 in 1993), and then, as we were inevitably rebuffed, we would ask for gradually smaller and smaller increments. Now, I only lasted for four call sessions, but I was amazed by the not inconsequential number of folks who stayed on the line and eventually would commit to $50 or $25. I didn't find out until much, much later that what the development office was using was straight out of psychological research.

This was first demonstrated on a college campus where two groups of students were asked about their openness to help out some troubled high school youth. In the first group, individuals were asked if they would go on an outing (as chaperones) with these teens on a two-hour trip to the zoo. In this condition, about one-eighth (16.7%) signed up. However, in the second group of participants, 50.0% of participants agreed to go to the zoo when asked this question (about the two-hour trip) second. The first question they were asked: would they commit to spending two YEARS as nonpaid counselors for the youth (i.e., the unreasonable request).

Low Ball

If you've ever bought a car, then you may wonder why it is that after you've committed to a price and feel like you've done a good job of staying within your budget (hopefully), you finally sit down to sign all of the paperwork and then get inundated with the following add-ons and fees like

- Rustproofing

- VIN etching

272

CHAPTER 10 MAKING DECISIONS: WHY WE BUY LOTTERY TICKETS

- Prepaid disposition fees
- Premium turn signal fluid
- Gap insurance
- Nitrogen-filled tires

OK, so only five of the six listed are actual products that they may try to sell you, but even though we all know the requests are coming, people still agree to these added costs. Why?

This technique, which is appropriately called the "low ball" technique, happens when a vendor or salesperson (the requestor) tries to sell you something at a low cost but then adds to it (using fees and other means) so that the final price is actually higher than what was originally agreed upon. In this situation, research has shown that the requestor (or seller) will be more successful if they start low and then increase the price than they would if they just advertised or tried to sell the product at the total price from the start. This technique works really well if one of the "hidden fees" (which of course can introduce all sorts of legal and ethical issues) isn't an alternative that you can live without (like the...ahem, premium turn signal fluid) but rather is necessary to completing the sale. Thus, in a case like that there is an unspoken commitment between the potential consumer and the seller that leads individuals to stick with making a purchase or engaging in a service even though the cost of doing so has increased.[31] The dark pattern that taps into this is drip pricing.

This has been demonstrated several ways (and takes place all the time with more than just automobile sales) but can also be about services. To this end, Guéguen and colleagues[32] found that if they asked individuals

---

[31] Burger, J. M., & Caputo, D. (2015). The low-ball compliance procedure: A meta-analysis. *Social Influence, 10*(4), 214–220.

[32] Guéguen, N., Pascual, A., & Dagot, L. (2002). Low-ball and compliance to a request: An application in a field setting. *Psychological Reports, 91*(1), 81–84.

273

CHAPTER 10 MAKING DECISIONS: WHY WE BUY LOTTERY TICKETS

on the street if they would hold the leash of a dog while they ran an errand that would take 30 minutes (which they then informed the individuals would actually take 45 minutes), they had many more individuals agree to do so then they did when they initially asked strangers to watch the animal for 45 minutes right from the beginning.

Something for Nothing

Finally, there is the "that's not all" (or the "I beat the system and got something for nothing") method that works if an offer is made and then (and this is key) before the individual can respond, either something additional is added to the deal (e.g., like doubling the amount or adding a t-shirt or some other "bonus") or the price is lowered slightly.

This particular phenomenon exploits the fact that individuals want to reciprocate generosity and so become more likely to accept the deal if the additional item (or new price) takes place sequentially (which gives the illusion of generosity on the part of the seller/requestor) rather than starting with the lower price or offering both the product and the additional bonus (e.g., the t-shirt) from the start.[33]

This phenomenon has also been demonstrated in a number of ways. For example, researchers held a bake sale (yep – you read that correctly) and were selling cupcakes and cookies. There were two conditions:

---

[33] Lee, S., Moon, S. I., & Feeley, T. H. (2019). The "that's-not-all" compliance-gaining technique: when does it work?. *Social Influence*, *14*(2), 25–39.

– **Condition 1**: When the researchers told individuals the price was $1 for either a cupcake or a bag of cookies but then were interrupted (by a phone call or another researcher), they would come back and tell the customer that in fact they could have the cupcake plus a bag of cookies for $1.00 (total).

– **Condition 2**: The researchers were up front that the cost of a cupcake and a bag of cookies together was $1.00

The researchers found that the student customers spent MORE money in Condition 1 (i.e., bought more cupcakes) than they did in Condition 2, even though they were actually getting a better deal in Condition 2 when they agreed to the sale.

Recap

- Most people will judge numbers – including prices – based on how they "feel" rather than what they actually mean or represent.

- These feelings result in common cognitive shortcuts that can skew decision-making, like a preference for frequencies over percentages, the gambler's fallacy, the hot-hand bias, the law of small numbers, and the leftmost digit bias.

- Conjunctive fallacies – the sense that if something is applicable to a large category it should also be applicable to combinations of categories – distort the likelihood of a feature being important or its likelihood to occur.

CHAPTER 10 MAKING DECISIONS: WHY WE BUY LOTTERY TICKETS

- Determining whether something is going to be profitable through cost–benefit analysis is trickier than it seems. Data are important but difficult to balance with less quantifiable factors.

- The impact of emotion on decision-making, which occurs through depletion of cognitive resources, cannot be overstated.

- Different factors like perceived or actual scarcity of products, as well as having too many options, all lead to poor decision-making.

- There are a number of techniques that have been demonstrated as having good efficacy in persuading individuals to engage in behaviors they may not otherwise do.

Before You Go...

Sometimes, scarcity can be used as a proxy for exclusivity. "PSL" fanatics (that's "Pumpkin Spice Lattes" for those of you not in the know) become excited at the end of every summer as they count down the days until leaves begin to fall and they can soothe their daily jitters with a hit of gourd-flavored bean water that will begin to be served again at their local Starbucks. The seasonality and predictability of PSLs have nothing, however, on what McDonald's has achieved with the seemingly random availability of their "restructured boneless pork patty" (that description just gets your mouth watering, doesn't it?): the McRib.[34] Though the McRib was

---

[34] https://www.today.com/food/restaurants/mcdonalds-mcrib-sauce-rcna181183

CHAPTER 10   MAKING DECISIONS: WHY WE BUY LOTTERY TICKETS

once a regular menu item, it was discontinued in 1985 due to lagging sales. After several years of outcry, however, the corporate McElders decided to re-release the sandwich for a limited time but to do so with reduced marketing and seemingly random locations where various franchises would carry the sandwich. This strategy has been incredibly successful for McDonald's, resulting in a loyal following (including a website to find out when and where the sandwich is available: `https://mcriblocator.com/`) and creating buzz[35] about what is, for all intents and purposes, a fairly humble offering.

---

[35] Deval, H., Mantel, S. P., Kardes, F. R., & Posavac, S. S. (2013). How naive theories drive opposing inferences from the same information. *Journal of Consumer Research, 39*(6), 1185–1201.

CHAPTER 11

Learning (and Making Mistakes)

JLR: So you know how I have three cats...

RB: Yes, and...

JLR: Well, one of them has started acting like a cat, and it is problematic.

RB: So, by "acting like a cat" do you mean "does whatever it wants, whenever it wants, however it wants"?

JLR: Exactly. See, she used to follow a training paradigm that I put into place perfectly – that is, at the end of the evening, I would say "bedtime for kittens," and she (Toni) and her siblings (Oscar and Emmy) would run into the basement and get in formation on a cat tower with three levels (one level per cat). As a reward, they would each get a treat.

RB: There's a lot to unpack there, but OK what is the problem?

JLR: Toni has decided that she doesn't like the flavor of treat that I'm using anymore.

RB: And so she is no longer engaging with the bedtime routine you've established?

JLR: Right. It turns out that when the reinforcer you are using is no longer a reinforcer, the behavior stops. And now I have to hunt her down, which is difficult as she is a tiny black cat and is really good at hiding. And I'm...

RB: Let me guess, "not so good at seeking"?

JLR: Exactly.

© Jerome L. Rekart and Rebecca Baker 2025
J. L. Rekart and R. Baker, *Designing for Human Intelligence in an Artificial Intelligence World*,
https://doi.org/10.1007/979-8-8688-1418-1_11

CHAPTER 11   LEARNING (AND MAKING MISTAKES)

"You'll Learn to Love it": Why Understanding How Learning Works Is Important for Design

Though the quote from the heading for this section isn't about design principles *per se*, as will become evident, the story behind it is valid for why understanding learning principles will result in easy to understand and navigate designs. Some years ago, one of us (Dr. Rekart) was shopping for a new car. He knew exactly what make, model, and features he wanted, as well as what colors he didn't. Given all of this, he visited a popular dealership (let's call it "Clueless Motors") with the hope that he would leave with a new set of wheels. After meeting and discussing his requirements with a salesperson (and going through all of the financial prequalifications, etc.), he was finally told that though the dealership had the desired year, make, and model on lot, the only vehicle that met all of those requirements was a color that was a nonstarter. When confronted with this, the not-so-savvy salesperson indicated that he (Dr. Rekart) would "learn to love it."

Let's just say that this was not the smartest sales tactic.

As this was some time ago, but not so long ago that access to the Internet wasn't at one's fingertips, after Dr. Rekart told the salesperson that he would need a few minutes to think about it, he took out his smartphone and searched the web for other local dealers. Within ten minutes, he found out that a vehicle that met all of his requirements was in stock fewer than 20 miles away (we'll call this other dealership "Gained the Next 15 Years of Business and No Fewer Than Six Vehicles Bought or Leased by Dr. Rekart Automotive"). He then promptly thanked the salesperson for his time and left that first dealership, never to return again.

Though we did cover persuasive techniques used by car salespeople (among others) in the last chapter, this anecdote isn't really about that. What it *is* about is the fact that having a mindset that customers or

CHAPTER 11 LEARNING (AND MAKING MISTAKES)

users will "learn to love" undesirable elements isn't just arrogant; it is detrimental to business.

Even when we open a well-trod site, our brain (as we'll discuss) is constantly searching for new stimuli to process. This fact should reinforce that learning is always taking place. What we do with that information and whether or not we store it was discussed in Chapter 9. What we're concerned with here is how to streamline and prejudice the information, concepts, and ideas that we want users to remember. How to stack the informational deck as it were.

When we think about design, best practices often focus on making experiences as easy as possible for end users. When users feel like a website is easy to learn, it increases engagement and feelings about the site.[1] Indeed, the notion of "self-directed learning" underlies much of how individuals are expected to interface with technology. When self-directed learning is facilitated, users are able to delight in features, whereas their frustration with nonoptimal experiences is often related to an inability to effectively (and quickly) learn how to use a given product in the manner that is needed (as is sometimes measured with proxies for frustration like "rage clicks").[2] This frustration can then take many forms, the least desirable being dissatisfaction with the product, which could lead to choosing a different product in the future.[3]

Or leaving the store and going to an entirely different one.

---

[1] Demangeot, C., & Broderick, A. J. (2016). Engaging customers during a website visit: a model of website customer engagement. *International Journal of Retail & Distribution Management, 44*(8), 814–839.

[2] Attard, C., Mountain, G., & Romano, D. M. (2016). Problem solving, confidence and frustration when carrying out familiar tasks on non-familiar mobile devices. *Computers in Human Behavior, 61*, 300–312.

[3] Wang, Q., & Shukla, P. (2013). Linking sources of consumer confusion to decision satisfaction: The role of choice goals. *Psychology & Marketing, 30*(4), 295–304.

281

CHAPTER 11   LEARNING (AND MAKING MISTAKES)

Associative Learning

We've probably all heard the story: some Russian scientist (i.e., Ivan Pavlov) trained a bunch of dogs to drool when a bell was rung.[4] Using the principles that would later become known as "classical" or "Pavlovian" conditioning, he elaborated on his initial observation that the salivation was caused by a bell that would ring whenever a door was opened to provide the dogs with their dinner of meat paste (which was even less appetizing than it sounds; it was essentially what you would get if you looked at a can of Spam and decided to make it soup). Pavlov showed that the bell, which was previously just a sound that had nothing to do with food (i.e., a neutral stimulus) became associated over time with something that would normally cause the dogs to drool, the meat paste. This form of associative learning was later demonstrated in any number of animals from sea slugs (who don't sound as yucky when they go by their Latin name of *Aplysia californica*) to honeybees (who sound cool in English and Latin, *Apis mellifera*) to dogs.

Later, studies by the American psychologist, John Watson, showed that emotional states in humans could also be evoked using this paradigm. In order to demonstrate this, Watson exposed an infant named "Little Albert" to a white rat. On its own, the rat did not bother Little Albert; however, by pairing the appearance of the white rat with deafening sounds that scared the infant, Watson was able to cause Little Albert to feel fear whenever he saw the rat. Thus, just like the dogs in Pavlov's studies, Little Albert learned to *associate* a previously neutral stimulus (in this case, the rat) with something else (frightening noises) such that on its own the rat would evoke fear in a way that it hadn't previously. The key to both

---

[4] Although there has been some revisionist writings suggesting that Pavlov did not use a bell, this has been thoroughly debunked and his usage of the term appeared in both print form and addresses he gave. Thomas, R.K. (1997). Correcting some Pavloviana regarding" Pavlov's bell" and Pavlov's" mugging". *The American Journal of Psychology, 110* (1), 115–125.

CHAPTER 11   LEARNING (AND MAKING MISTAKES)

of these studies (and all others like them that pair neutral stimuli with natural stimuli that evoke responses) is that both of the behaviors that were observed were involuntary (i.e., the salivation of the dogs and the crying and fear of Little Albert). This latter point is one that often gets misremembered or forgotten by students of psychology. And that gets distorted when people talk about the impact of various stimuli on human behavior.

This misunderstanding may be due, in part, to the fact that many modern advertising and market research techniques were actually first put into place by the same Watson we just described torturing a small child with a rodent.[5] The feelings that Watson wanted to instill in consumers were, as with Little Albert, involuntary. For example, to build unconscious positive inclinations toward products, which would then increase the likelihood that they would be purchased.

In order to accomplish this, one must pair a previously neutral stimulus with one that engenders a given response on its own, just as Pavlov did with the bell and the meat paste. Based on many individuals' understanding of this phenomenon, we are bombarded today with positive images and sounds (i.e., think music) that marketers hope will become associated with their products.

Now, if this were to work via a classical conditioning mechanism, that would mean that after a number of pairings, the advertised product alone would engender positive feelings. On the surface, this seems like a great way to transfer positivity from one stimulus to another. Think about the number of Hollywood A-listers and professional athletes whom you associate with various products. Why is that – a desire to have some transference of positivity from them to what they are hawking. Indeed, in the United States alone, celebrities endorse as many as one-quarter

---

[5] Buckley, K. W. (1982). The selling of a psychologist: John Broadus Watson and the application of behavioral techniques to advertising. *Journal of the History of the Behavioral Sciences, 18*(3), 207–221.

CHAPTER 11   LEARNING (AND MAKING MISTAKES)

of all products advertised.[6] With broad smiles and cultural relevance, individuals believed to be viewed positively attest to the value, efficacy, and virtues of various products. But is this classical conditioning? And, regardless, is it even effective?

The answer to the first question is, decidedly, "no."[7] There are a number of theories about why some forms of advertising work and others do not. For example, images and sounds can evoke emotions on their own. The imagery of a cheeseburger on screen has a decent chance of making someone who is hungry salivate and *think of food*[8] regardless if it is being held by the quarterback of a successful professional football team or a starlet dressed to the nines (or vice versa). But, this stimulus–response pairing is not one that produces thoughts one may have about either of the two humans and extends it to the food. Nor, presumably, does it take feelings about the food and extend it to the humans!

The answer to the second question is more of a qualified "it depends." Indeed, some celebrity endorsements do in fact impact sales, but there are a number of parameters that impact this,[9] such as

- The credibility of the endorser (increases sales)

- How much the endorser can be believed to actually use, know, or be associated with a given product (increases, if believed)

---

[6] Erdogan, B. Z. (1999). Celebrity endorsement: A literature review. *Journal of marketing management, 15*(4), 291–314.

[7] Pornpitakpan, C. (2012). A critical review of classical conditioning effects on consumer behavior. *Australasian Marketing Journal, 20*(4), 282–296.

[8] Folkvord, F., Anschütz, D. J., Boyland, E., Kelly, B., & Buijzen, M. (2016). Food advertising and eating behavior in children. *Current Opinion in Behavioral Sciences, 9*, 26–31.

[9] Knoll, J., & Matthes, J. (2017). The effectiveness of celebrity endorsements: a meta-analysis. Journal of the academy of marketing science, 45, 55–75.

CHAPTER 11   LEARNING (AND MAKING MISTAKES)

- How many other products the individual endorses (decreases with the number of products)

- The level of perceived attractiveness of the endorser (we aren't saying we like that this is the case, but it is what the data show, and as you might expect, as attractiveness of the endorser goes up, so do sales)

In an expansive meta-analysis that examined the efficacy of celebrity endorsers, it was actually found that though they did have an overall positive impact on products (across dozens of studies that were examined in the aggregate), their overall impact was less than "awards" or "seals of approval," probably because the latter are believed to convey the findings of unbiased evaluators.

So what does all of this mean? In a nutshell, it means that there are ways to positively impact the perception of one's designs or product, but that it (1) is not due to classical conditioning and (2) the reasons why it may or may not work for a given consumer or user will be complex and likely well beyond the scope of any user or market research you may conduct.

The Carrot Is Mightier Than the Stick

Unlike classical conditioning, another form of associative learning called "instrumental," "operant," or "Skinnerian"[10] conditioning actually changes the behavior of individuals who have been conditioned. What's more, in instrumental conditioning paradigms, individuals are aware and make

---

[10] Although technically it should be called "Thorndikian" as Edward Thorndike was the first to characterize it.

CHAPTER 11 LEARNING (AND MAKING MISTAKES)

choices about how and when to behave (i.e., it is voluntary and under their control). This means that the individual whose behavior is conditioned makes a choice to act in that particular way.[11]

Instrumental conditioning follows what is known as the law of effect, which was first demonstrated by Edward Thorndike.[12] The law of effect is essentially the principle that those behaviors that achieve or result in favorable outcomes will be more likely to be repeated (and conversely, those that result in unfavorable outcomes will be less likely).

The result of conditioning is thus to facilitate preference for one behavior over another. This is a subtle but important distinction about instrumental conditioning. Whereas with classical conditioning, the dogs couldn't control when (or if) they started to drool when they heard a bell (or saw the white lab coats of the experimenters), with instrumental conditioning, which is how dogs (and children) are housetrained, the conditioned behavior (relieving oneself in an appropriate location) occurs because the dogs have come to associate the desired (by the owner) behavioral choice with a situation that is preferable to them in the form of affection or treats of some sort.

These preferable outcomes are called "appetitive" because they draw the learner in. Conversely, contingencies that are undesirable are termed "aversive" because they push the learner away. Thus, reinforcers are appetitive and punishment is aversive. For instrumental conditioning to occur, a contingency (either appetitive or aversive) needs to occur either when an animal (including humans) behave in a particular way or soon after the behavior has taken place.

Although dogs (and humans) can learn which behaviors are preferred by those around them through punishment, decades of studies show this to be much less effective than rewards. Researchers refer to these

---

[11] Staddon, J. E., & Cerutti, D. T. (2003). Operant conditioning. *Annual review of psychology*, 54(1), 115–144.

[12] Thorndike, E. L. (1933). A proof of the law of effect. *Science*, 77(1989), 173–175.

CHAPTER 11   LEARNING (AND MAKING MISTAKES)

contingencies as reinforcers (appetitive) and punishment (aversive), and there are two types of each depending upon whether something is added (i.e., positive) or subtracted (i.e., negative), for a total of four instrumental conditioning contingencies. This means that both positive punishment and negative reinforcement are possible, with the latter being preferred by the learner and teachers, designers, pet owners, and parents who are "in-the-know." Let's say that you're a parent who wants your teenage son to do the dishes. There are four different approaches you could take to facilitating this:

- **Positive Reinforcement**: Every time your son loads the dishwasher, he gets $1 added to his allowance. (addition of something appetitive)

- **Negative Reinforcement**: Every time your son loads the dishwasher, he doesn't have to set the table the following dinner. (subtraction of something aversive)

- **Positive Punishment**: Every time your son *doesn't* load the dishwasher, you yell at him. (addition of something aversive)

- **Negative Punishment**: Every time your son doesn't load the dishwasher, you reduce the amount of time he can spend playing video games by ten minutes. (subtraction of something appetitive)

As you can see from these examples, there are different ways to reinforce a behavior. Again, what research has shown is that the first two (i.e., reinforcing a behavior) are more effective than the latter two. You (hopefully) also noted that negative reinforcement does not mean "reinforcing something with an aversive behavior," even though this is how negative reinforcement is often used colloquially (e.g., drawing attention to someone misbehaving in a class isn't "negative reinforcement," but rather positive reinforcement, if the attention is desired. If it isn't, then it would be positive punishment).

287

CHAPTER 11   LEARNING (AND MAKING MISTAKES)

Designers can incorporate these principles in many different ways. Some of the most obvious ones are

- Badges

- Rewards

- Likes/thumbs up (these provide positive social reinforcement)

But it is important to recognize that there are other forms of positively reinforcing contingencies that can also occur, such as by providing opportunities for users to personalize elements of their experience. Indeed, responsiveness of designs to personal inclinations or needs increases the speed with which users learn how to navigate sites, which in turn fosters engagement.[13]

The strength of the learning that takes place – as measured by the choice to produce the reinforced (or refrain from the punished) behavior – is impacted by the frequency with which the behavioral contingency (i.e., reinforcement or punishment) occurs.[14] The fastest way to mold behavior is by using a fixed ratio. With a fixed ratio "schedule of reinforcement," a behavior is reinforced every Nth time it occurs. Thus, rewarding your son every time he washes the dishes or after every seventh night he does so are both fixed ratios (1:1 and 7:1, respectively). A recent study examined the potential of earning points (to be used within the app) or a leaderboard (seen as having positive value by the participants) as reinforcers of app engagement (i.e., the behavior of choice was opening the app and using it).[15]

---

[13] Ferretti, S., Mirri, S., Prandi, C., & Salomoni, P. (2017). On personalizing web content through reinforcement learning. *Universal Access in the Information Society, 16*, 395–410.

[14] Morgan, D. L. (2010). Schedules of reinforcement at 50: a retrospective appreciation. *The Psychological Record, 60*, 151–172.

[15] Garaialde, D., Cox, A. L., & Cowan, B. R. (2021). Designing gamified rewards to encourage repeated app selection: Effect of reward placement. *International Journal of Human-Computer Studies, 153*, 102661.

288

CHAPTER 11   LEARNING (AND MAKING MISTAKES)

Not surprisingly, the authors found that the greatest impact on app behavior was seen when the reward (either one) was presented to the user as soon as they logged on, with less efficacy observed the longer a user had to wait before being "rewarded" with either of the two options.

Although a fixed ratio schedule of reinforcement results in the fastest learning, it is also the weakest way to keep someone performing the desired behavior. Remember that what is learned is "behavior = reward"; therefore, when the latter part of that equation is removed, the learning undergoes "extinction" (i.e., is extinguished). Now, as the ratio between the behavior and the reinforcement increases (like going from a reward every night to every week), the behavior becomes more resistant to extinction, though this isn't because the learning is any stronger; it is just that there is a built-in interval when no reinforcement is expected. However, once the reward isn't provided, new learning starts to take place.

On the other hand, the most resistant schedules of reinforcement (i.e., those that result in the persistence of the desired behavior) involve varying either the time between the behavior and reinforcement (variable interval) or the number of occurrences between them (variable ratio). These parameters do take much longer for learning to take place, which, again, makes sense as the association between them (i.e., the underlying pattern) requires more occurrences before it becomes evident to the learner. This form of reinforcement is why certain addictions, particularly those that involve gambling, are so difficult to treat. It also explains why the most effective implementations of "gamification" utilize partial reinforcement schedules that intersperse both fixed and variable ratios of reinforcement.[16]

---

[16] Richter, G., Raban, D. R., & Rafaeli, S. (2015). *Studying gamification: The effect of rewards and incentives on motivation* (pp. 21–46). Springer International Publishing.

CHAPTER 11   LEARNING (AND MAKING MISTAKES)

TAKING OUT THE TRASH

This principle has driven our chore system at home, with marked success. We provide a mix of immediate and delayed positive reinforcements in the form of points (once a chore is complete, you get a certain number of points that show up on your whiteboard) that can be turned into cash weekly for necessary regular chores (cleaning up the yard from dog messes, etc.). For small chores, like setting the table, we give a pass for another small chore like putting away the leftovers. We couple that with a negative punishment (removal of electronic devices if the chores are not completed by a set deadline). While it now seems very straightforward how the positive/negative reinforcement variables work, it was definitely a process setting it up with a great deal of trial and error (and more than a few eye rolls). The greatest key to our success was that we involved the users of the system in the election of reinforcements. That is, we talked to the kids about what would be effective and ineffective methods of positive and negative reinforcement (which turned out to be far different than we expected – and have evolved over time. When they were younger, they valued privileges like going to the zoo over money, but as they got older, they valued money and time more). As a result, we were able to fine-tune the system to be effective for our user base. Research for the win! (and for getting the trash to the curb!)

"Learning Scientists Swear by This One Trick..."

We opened up Chapter 9 by defining memory and in so doing mentioned that it could be a "skill, fact, concept, or..." and then provided a rather absurd example of what it could be (and even provided an illustration to drive home the point). Do you remember what came after the "or"?

290

CHAPTER 11   LEARNING (AND MAKING MISTAKES)

We'll give you a hint: the image that we described was one of a particularly dangerous animal looking rather jaunty – do you remember it now?

Chances are good that in fact, even before reading the hint about the "dangerous animal" (which they really are[17]...and not just to marbles), you recalled seeing a line drawing of a hippopotamus wearing a top hat. Indeed, we (perhaps sneakily) included that absurd example with the express purpose of evoking it later in this book because we knew there was something about it that would grab your attention in a way that a rather ho-hum example might not: its novelty.

The impact of originality on attention is a fairly well-understood phenomenon[18] that plays out in demonstrable ways on a daily basis.

"You won't believe what happened next!"

"Top expert reveals the biggest secret to making $$$$."

"Use this one weird trick to live longer."

Clickbait, which we encounter along the margins of our browsing lives, is nothing more than an attempt to draw us in with something that is titillating or outrageous enough to get us to navigate to it (and thus draw our eyeballs and, importantly, our credit cards, toward whatever is being hocked). Indeed, what has been deemed our "thirst for novelty"[19] leads us

---

[17] van Houdt, S., & Traill, L. W. (2022). A synthesis of human conflict with an African megaherbivore; the common hippopotamus. *Frontiers in Conservation Science, 3,* 954722.

[18] Schomaker, J., & Meeter, M. (2015). Short-and long-lasting consequences of novelty, deviance and surprise on brain and cognition. *Neuroscience & Biobehavioral Reviews, 55,* 268–279.

[19] Lorenz-Spreen, P., Lewandowsky, S., Sunstein, C. R., & Hertwig, R. (2020). How behavioural sciences can promote truth, autonomy and democratic discourse online. *Nature human behaviour, 4*(11), 1102–1109.

CHAPTER 11 LEARNING (AND MAKING MISTAKES)

to spend increasingly more time online, where we in turn click on more sites, reels, and links, with the end result that most of it is actually not, in fact, novel. Just endless variations on themes.

However, what if what happened next, the biggest secret, or the weird trick really were something unbelievable, too good to be true, or in fact, quite weird? If any of those things were true – meaning that the new information were in fact original – then research has shown that the likelihood that we would remember this new information would in fact be quite high[20] – even if it isn't something that is of use to us, like a well-dressed African megaherbivore.[21]

This attraction to, and proclivity to store, novel information is due in part to the role of our favorite brain region (or at least it is Dr. Rekart's): the hippocampus. The hippocampus has been implicated in the detection and preferential storage of novel or extraordinary information for quite some time.[22] If we go back to our earlier discussion of the role of the hippocampus across the centuries (again, Chapter 9), then we should be reminded that from an evolutionary point of view, it evolved to help us remember information that was critical to our survival and that of the species. Knowing where we would be able to find shelter, food, water, and others like us (i.e., family and friendly companions) was advantageous to the longevity of our ancestors, and those who were able to stay alive longer had more opportunities to pass on their genes. But just storing old information (e.g., where the apple trees are located along the stream) wouldn't have been enough as the only constant in life is change. Thus, where a new predator was seen (down by the rock cropping), what preceded the untimely death of an associate of ours (that weird green berry

---

[20] Rekart, J. L. (2013). *The cognitive classroom: Using brain and cognitive science to optimize student success*. R&L Education.

[21] A hippopotamus with a top hat.

[22] Knight, R. T. (1996). Contribution of human hippocampal region to novelty detection. *Nature, 383*(6597), 256–259.

CHAPTER 11   LEARNING (AND MAKING MISTAKES)

we hadn't seen before), etc. would also be important information to store and also was a contributor to the continued existence and development of our species (Go Team!).

In the modern era, we can now exploit this evolutionary legacy in a number of different ways. For example:

- By using color differences to draw the eye to particular locations

- Through unique characters or "mascots," like Duo the Owl (from the Duolingo app), which can then be modified to tweak engagement[23]

- Through occasional refreshes and redesigns that move the user away from familiar routines on the site to active exploration[24]

It is best to recognize that novelty should be implemented within the constraints of good taste. That is to say that though the relationship between attention, memory, and novelty may be linear, this is not the case for how pleasing something is considered to be (i.e., aesthetic appeal). Indeed, research has shown that moderate levels of novelty are considered the most aesthetically pleasing, with very familiar designs of items such as website or chairs, considered to be "meh," and those that are the most new as being less pleasing to the eye or beautiful than less novel iterations.[25] Put differently, novelty for novelty's sake is never a good idea.

---

[23] https://www.forbes.com/sites/lesliekatz/2024/09/06/why-the-duolingo-owl-suddenly-looks-so-sick/

[24] Barnes, S. J., & Vidgen, R. T. (2001). Assessing the effect of a Web site redesign initiative: An SME case study. *International Journal of Management Literature, 1*(1), 113–126.

[25] Hung, W. K., & Chen, L. L. (2012). Effects of novelty and its dimensions on aesthetic preference in product design. *International Journal of Design, 6*(2), 81–90.

CHAPTER 11   LEARNING (AND MAKING MISTAKES)

Why We Suck at Remembering *Some* Things (and What We Can Do About It)

According to his 1885 work, *Memory: A Contribution to Experimental Psychology*, Hermann Ebbinghaus worked out the idea that one should "work smarter, not harder."[26] On the surface, this may not seem evident based on his approach (nor should one think that anywhere in the volume he actually uses that phrase) which involved thousands of pieces of written information, some of which looked like this:

ILT

KLA

OMB

Though these may resemble products one could purchase from IKEA, they were actually nonsense syllables that he constructed. He used nonsense syllables because he was trying to remove the influence of any prior learning that he may have had that could help him to remember strings of these nonsense syllables (e.g., BAT + MAN = BATMAN). By just introducing himself to the nonsense, he hoped to identify how one could store new information efficiently.

Seeing as we are writing about his findings some 150 years later, the title of his book, which may have seemed an act of hubris at the time, was proven to be correct. His most well-known and still germane *contribution* was showing that he could reduce the amount of time he needed to study a list of nonsense syllables by spreading it out over several days (called "spaced" learning) rather than trying to learn everything in one sitting

---

[26] Ebbinghaus, H. (1913/1885). Memory: A contribution to experimental psychology (H. A. Ruger & C. E. Bussenius, trans.). New York: Teachers College, Columbia University.

CHAPTER 11   LEARNING (AND MAKING MISTAKES)

(called "massed" learning or, in the vernacular of students everywhere, "cramming"). This technique, which is called spaced repetition learning (SRL), involves spacing out information across several sessions in order to facilitate entry into long-term memory. This technique is effective for much more than just verbal tasks (above and beyond nonsense syllables, that is),[27] with evidence showing its efficacy in the learning of everything from elementary school scientific concepts[28] to college-level calculus.[29] In addition to being a great technique for the classroom, SRL is also an effective way for adults to retain information needed on the job.[30]

The efficacy of this technique relies not in the repetition per se, but in the spacing of information across multiple sessions. At the neurological level, it may be helpful to think about neural pathways in the same way that bodybuilders think of their physiques. For a bodybuilder to gain muscle mass, they must communicate to their bodies that the amount of mass they currently possess is insufficient. How do they do this? By repeatedly exposing their muscles to loads that cause their muscles to fatigue and stress (and quite literally tear at the microscopic level), the athletes are able to "tell" their bodies that it is time to step it up and make more muscle. Similarly, when we come back to information that we don't

---

[27] Cepeda, N. J., Pashler, H., Vul, E., Wixted, J. T., & Rohrer, D. (2006). Distributed practice in verbal recall tasks: A review and quantitative synthesis. *Psychological bulletin, 132*(3), 354.

[28] Gluckman, M., Vlach, H. A., & Sandhofer, C. M. (2014). Spacing simultaneously promotes multiple forms of learning in children's science curriculum. *Applied Cognitive Psychology, 28*(2), 266–273.

[29] Hopkins, R. F., Lyle, K. B., Hieb, J. L., & Ralston, P. A. (2016). Spaced retrieval practice increases college students' short-and long-term retention of mathematics knowledge. *Educational Psychology Review, 28*, 853–873.

[30] Kim, A. S. N., Wong-Kee-You, A. M. B., Wiseheart, M., & Rosenbaum, R. S. (2019). The spacing effect stands up to big data. *Behavior Research Methods, 51*, 1485–1497.

CHAPTER 11   LEARNING (AND MAKING MISTAKES)

have in our long-term memories (yet) during multiple sessions, we are "telling" our brains that this information is important enough to create new pathways in the brain.[31]

Creating new neural pathways, like building muscle mass, requires a large store of cellular energy and dedication of resources (e.g., protein, organelles, scaffolding molecules, etc.), so neither our muscular system nor our brains want to just undertake such endeavors without sufficient reason to do so. Spaced repetition learning is the way via which we convey the value of the pathways to be built.

Although the spacing in SRL refers to the temporal domain, a variation of this phenomenon called interleaving[32] shows there is utility in also spacing out (rather than "blocking") categories of information so that it is varied. Thus, if you had three decks of flash cards, one each for upcoming finals in economics, psychology, and finance, you would optimize studying by shuffling all three together and reviewing a different combined deck at different points over a week, thus spacing them out across both temporal and categoric domains.

Though certainly many modern learning platforms, like Duolingo, integrate spaced repetition learning into their models, its implementation should not be seen as only being the purview of education. If one of the hallmarks of good design is consistency, then you can think of SRL (with interleaving) as justification for surfacing information that you need or want your users to learn and retain. This could be through pop-up nudges that are strategically timed and placed, placement of brand messaging, etc. The key is that when a user is using your product, you have

---

[31] Rekart, J. L., Sandoval, C. J., Bermudez-Rattoni, F., & Routtenberg, A. (2007). Remodeling of hippocampal mossy fibers is selectively induced seven days after the acquisition of a spatial but not a cued reference memory task. *Learning & Memory*, *14*(6), 416–421.

[32] Sana, F., & Yan, V. X. (2022). Interleaving retrieval practice promotes science learning. *Psychological Science*, *33*(5), 782–788.

CHAPTER 11   LEARNING (AND MAKING MISTAKES)

an opportunity to reinforce and reintroduce brand- and product-specific messaging that provides value that will make their experience better, more manageable, and ultimately make them want to return.

Failing Forward

There is a myth that Albert Einstein made the following assertion:

Everyone is a genius. But if you judge a fish by its ability to climb a tree, it will live its whole life believing that it is stupid.

Though perpetuated by memes and numerous shares on the Internet, there is actually no proof that the German physicist ever uttered those words.[33] However, despite the suspect attribution, the sentiment does have some applicability to what we are discussing. As with evaluating the intelligence of the fish, researchers must think about how to evaluate learning in ways that are meaningful and possible and reflect the capabilities of their subjects. Furthermore and more germane to the topic of this book, the "learning-performance distinction" highlights the fact that learning can take place even when it isn't explicitly observed (i.e., performed).[34] As anyone who has ever come out of a test thinking that they knew much more than they were able to demonstrate would attest, this is a common phenomenon in daily life. It is also reflective of the fact that even once we are no longer actively *consciously* processing information, our minds may be continuing to do so in the background. This seemingly "offline" processing is responsible for the moments of epiphany that

---

[33] O'Toole, G. (2017). *Hemingway didn't say that: The truth behind familiar quotations*. little a publishing.

[34] Soderstrom, N. C., & Bjork, R. A. (2015). Learning versus performance: An integrative review. *Perspectives on Psychological Science, 10*(2), 176–199.

297

CHAPTER 11   LEARNING (AND MAKING MISTAKES)

we may experience (see the bonus story in Chapter 6), as well as some increases in problem-solving that are seen not after short (e.g., minutes), but longer delays (e.g., weeks).

When learning appears to be initially lacking, but then manifests after a delay, it is called "productive failure." Productive failure is often built into educational settings *before* teachers provide instruction on how to solve complex problems, like story problems in mathematics. In the earliest studies of this phenomenon, it was shown that although the students who weren't given any formal instruction (PF) were at an initial disadvantage when trying to solve the problems relative to a group that read examples (productive success; PS) about how to solve them, after the instruction, the PF group performed significantly better with novel (though related) problems than the PS group.[35]

Given that most of a user's interactions with a product will involve self-directed learning, it might seem contradictory to introduce desirable difficulties in order to facilitate productive failure. Indeed, why would someone want to deliberately create opportunities for users to "flail about" given that this is antithetical to the designer's mandate to simplify. The answer is – as with many of the recommendations in this chapter – to be judicious about where and when to use these techniques. When it comes to productive failure, luckily studies have shown that the best way to integrate them into designs is regarding instructions. Thus, if there are elements that require instruction or could be slightly more complicated, it makes sense (and, yes, studies have validated this) to build an opportunity for your user to explore and figure out where to go before you hit them with the "here's the tutorial link."[36]

---

[35] Kapur, M. (2008). Productive failure. *Cognition and instruction, 26*(3), 379–424.

[36] Kapur, M. (2016). Examining productive failure, productive success, unproductive failure, and unproductive success in learning. *Educational Psychologist, 51*(2), 289–299.

298

CHAPTER 11   LEARNING (AND MAKING MISTAKES)

Recap

- When someone uses a product, they are learning about it and how it fits into their lives and meets their needs (or not).

- Although it is helpful to convey or advertise products in a manner that evokes positive feelings, this does not occur, contrary to popular belief, via classical conditioning.

- Thorndike's law of effect is important because it holds that any parameter that increases a behavior that a designer wants to occur should be reinforced within their designs.

- Learning can be quickly and easily facilitated using various principles, such as the implementation of novelty and spacing information by category and across time by repeating cues.

- Failure can be productive in the sense that not only may it not be an accurate indicator of learning but that it could also facilitate a deeper understanding when properly implemented.

Before You Go...

The year was 2005 and a fresh-faced (OK, maybe not) first-year graduate student had been assigned to be the teaching assistant (TA) for his graduate mentor (i.e., the professor under whom he would be conducting the research for his master's and doctoral degrees) whom we will refer to as "Professor AR." The course in question was a 200-level undergraduate course on the

299

CHAPTER 11 LEARNING (AND MAKING MISTAKES)

biological workings of the brain and its relation to behavior: Introduction to Neuroscience. Professor AR, who at the time had been tenured longer than this graduate student had been alive, decided that he was going to "mix things up" and would, with the graduate student's assistance, teach the course backward. Yep. Start with the material normally only covered in the last week and end the course – right before the final – with the most basic information and definitions. In practice, this meant that the textbook was no longer applicable as it had the material and concepts structured to build upon one another over the course of the term: early chapters talked about what a neuron was and how it communicated with other neurons, while later ones described how that communication could be disrupted in mental disorders or would be responsible for the action of drugs across various locations in the brain. Unfortunately, the protestations of the young TA that this maybe wasn't an approach that would be "appreciated" by the students fell on deaf ears. And thus during that semester, 35 unsuspecting undergraduates were taught the workings of the nervous system in reverse. Now, this experiment in pedagogical technique went about as well as you are probably thinking it did. Although Professor AR had a merry old time lecturing about all of his favorite things in an entirely different order, the confusion, frustration, and outright anger expressed by the students provided quite the baptism by fire to college-level teaching for the TA who spent the term deciphering lectures, holding ad hoc discussion sections to get students up-to-speed (quickly), and producing study guides to help them make sense of the material being presented by the esteemed professor. At term's end (and once the dust had settled), Professor AR was nice enough to share the course evaluations with the TA, who was mentioned in glowing terms as the "savior" of the course. This particular anecdote is shared here for two reasons: (1) it started one of your authors on his journey toward thinking about practical applications of cognitive science principles, and (2) it shows that even for a learning scientist of global recognition, sometimes the best laid plans don't work out the way they're intended (he was attempting an "interleaving, spaced training melange" as he put it).

CHAPTER 12

Business, Research, and Design Relationships: It's Complicated

RB: I was part of a roundtable discussion about how to do UX research effectively for companies, and the question came up: How do you use research to show the value of UX?

JLR: What did you tell them?

RB: Like a Facebook relationship status, it's complicated.

JLR: <snort>

RB: But, seriously, showing the value of research depends on a lot of things. Perhaps the most important thing is making sure your company understands that you do not really measure user experience. You measure the results of user experience.

JLR: Precisely. We measure outcomes, which means you need to understand what you expect to happen before you measure whether or not it (whatever it is) happened. Problems before solutions, success definition before measurement.

RB: Journey before destination (IYKYK).

© Jerome L. Rekart and Rebecca Baker 2025
J. L. Rekart and R. Baker, *Designing for Human Intelligence in an Artificial Intelligence World*,
https://doi.org/10.1007/979-8-8688-1418-1_12

CHAPTER 12   BUSINESS, RESEARCH, AND DESIGN RELATIONSHIPS: IT'S COMPLICATED

Design Thinking: No Pipe Cleaners Version

Design thinking[1] is a hot topic for many companies. Entire consultancies have been built around facilitating design thinking workshops complete with sticky notes, pipe cleaners, markers, molding clay, and other craft items normally found in an elementary school's art teacher's closet. But what exactly is design thinking?

Popularized by IDEO in the 1990s,[2] simply put, design thinking is a framework used to promote solution-focused thinking. Leveraging some of the techniques used in games, design thinking creates a safe, collaborative environment that cultivates a connection between the participants and the user. Working together to define a problem through research, then ideating and testing a solution creates a shared experience with the user (as you define their situation and challenges) and with your fellow designers. Meant to be done in collaboration with users, design thinking follows five stages: Empathize, Define, Ideate, Prototype, and Test.

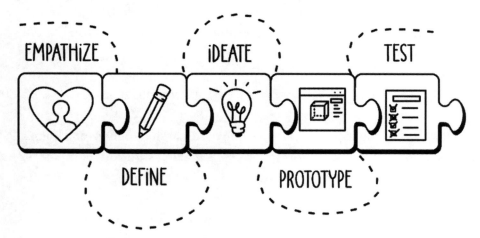

---

[1] https://en.wikipedia.org/wiki/Design_thinking
[2] https://hbr.org/2008/06/design-thinking

CHAPTER 12   BUSINESS, RESEARCH, AND DESIGN RELATIONSHIPS: IT'S COMPLICATED

Empathize – with Data

Many companies are big fans of data. "We are a data-driven company!" is the battle cry of many a CEO. However, the average employee will tell you behind their hands "We are a driven company…with a lot of data." Why do we have this disconnect? Well, as we pointed out in earlier chapters, humans have a lot of cognitive biases. As such, it is easier to ignore the math and "go with our gut" (spoiler alert – your gut sucks at making logical decisions – why would you trust a tube to conduct any serious business?). However, to understand our users' needs and context, we must set this tendency aside and take the time to listen to our data.

Why Eating Ice Cream Causes Shark Attacks (or NOT)

There is a strong, well-articulated statistical relationship between the amount of ice cream that is eaten and the number of shark attacks and drownings that occur. So what is going on here? Should the surgeon general place a warning label on pints of rocky road and butter brickle? Are Ben and Jerry actually super villains plotting to control the world's population through oceanic interventions?[3] Is this why your mother always told you to wait a half-hour before swimming after a meal?

Nah. The answer to the first is "no"; to the second, "probably not"; and to the third is "maybe, but there is no scientific reason you can't swim after eating."

---

[3] To be clear, we are in no way indicating that we have uncovered a plot by Ben and Jerry's involving sharks with laser beams on their heads or any other volcano-lair-based activities. This is absolutely fiction and should not trigger any attempts to silence the authors with overly complicated death scenarios accompanied by monologues. Please.

303

CHAPTER 12   BUSINESS, RESEARCH, AND DESIGN RELATIONSHIPS: IT'S COMPLICATED

The issue isn't that somehow eating ice cream and shark attacks are related but that we automatically assume that relationships are causal ones. If you go back and look at what we wrote, you'll see that all that we indicated was that there was a "strong, well-articulated statistical relationship"; we didn't write that one caused the other. So what is going on?

This example, which introductory statistics professors love to use, illustrates the importance of considering so-called "third variables" and reinforces that correlation does not necessitate causation. Any ideas what the third variable is for this scenario (i.e., shark attacks and ice cream)?

If you said something along the lines of temperature, the seasons (e.g., autumn, spring), or time of year, you'd be correct. There are logical, well-known relationships between the temperature outside (which in the summer is warmer than the winter) and the number of people who partake in swimming outside, where shark attacks are most likely to occur. Thus,

CHAPTER 12 BUSINESS, RESEARCH, AND DESIGN RELATIONSHIPS: IT'S COMPLICATED

there is a correlation between air temperature, the likelihood of eating ice cream (at least in Dr. Rekart's neighborhood, the ice cream truck doesn't make rounds when there is snow on the ground), and the likelihood that one may swim. The more one swims, the more their chances of falling afoul of a toothy ocean predator increases.

Notice that we've been careful to indicate that all of these relationships are correlations. Even between air temperature and shark attacks, we haven't stated that "X" causes "Y." This is because that would mean that somehow the air temperature directly influences the probability of the outcome in question. Although we could generate some convoluted explanation that equates increases in air temperature concomitantly increasing water temperature, which in turn increases the activity of water molecules, blah, blah, blah…that still wouldn't be accurate.

Thus, the take-home messages are that when you are considering relationships among variables in your data, you need to:

- Not draw causal inferences from correlations (an experiment is the only methodology that allows for this). Put another way, we shouldn't conclude that two things are related just because they both increase/ decrease at the same time.

- Consider third variables that may actually be more directly related to the phenomena or results you are seeing. That is, think about things outside your data set that might be impacting your results.

305

CHAPTER 12   BUSINESS, RESEARCH, AND DESIGN RELATIONSHIPS: IT'S COMPLICATED

Data Trends: What Got You Here Won't (Necessarily) Get You There

Another issue that can occur with correlations is making predictions that exceed the bounds of your data. What does that mean? Well, this is like trying to predict your sales will continue to trend up at 2% per quarter because the last two years of data shows sales trending at 2% a quarter.

IT'S COMPLICATED

This type of mistake happens when trend data are used to make predictions about the behavior of two different groups into the future based solely on past actions. And although this is a common practice, modern techniques overcome many statistical problems by aggregating large bodies of data and using multiple variables to overcome and control for third-variable problems that may be inherent within the data when simple relationships are examined. That is, we have ways to work around it, but you have to know what you're doing and be careful to avoid the ice cream/shark problem.

Let's take a look at an example. In 1992, a brief article was published in the prestigious journal, *Nature,* with the title[4]: "Will women soon outrun men?" This article plotted data for Olympic running events for both men and women from the early 1900s up through 1988. The trends observed by the authors were (recreated in the accompanying figure) as follows:

- Over the course of the 20th century, both men (circles) and women (triangles) recorded gradually faster times (i.e., their average speed increased) year over year (i.e., runners for both sexes kept breaking records with faster times).

---

[4] Whipp, B. J., & Ward, S. (1992). Will women soon outrun men?. *Nature,* *355*(6355), 25.

306

- The gains made by men over the 20th century were more incremental than those made by women (i.e., the slope of the curve for women's fastest times was steeper than those of men's).

- The slope of the curve for women in the marathon was the steepest.

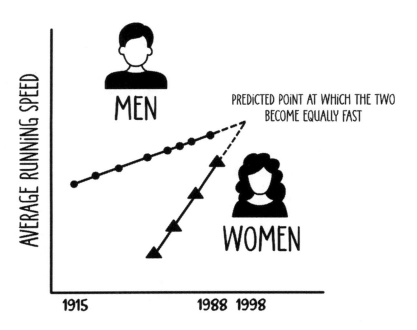

Now, none of the reported, actual data (at the time) were inaccurate or represented misinterpretations whatsoever: the data were what the data were (represented by the solid lines in the figure). Where things went awry is that the researchers then projected the two curves into the future (see dashed lines in the figure) and concluded that men and women marathoners would be running equally fast in 1998.

Is this what happened?

CHAPTER 12   BUSINESS, RESEARCH, AND DESIGN RELATIONSHIPS: IT'S COMPLICATED

Even without a Google search you may think "that doesn't sound right," and you'd be correct. As of the last Olympics (which took place in Paris during the summer of 2024), the gold medal winners from both sexes posted the following times for their respective events:

- Men: 2:06:26

- Women: 2:22:55

And if you think those are fast times to run 26.2 miles (they are!), then check out the world records as of early 2025 (years and locations in parentheses):

- Men: 2:00:35 (2023; Chicago Marathon)

- Women: 2:09:56 (2024; Chicago Marathon)

As you can see from both sets of current data, we are still not at a place where women are running as fast as men in distance (or any running) events – even though over a quarter of a century has passed from when the model shown above suggested we would. Note that this does NOT mean that there aren't women who are faster than some men (there are many). It means that the supposition of the authors in that 1992 paper that *among elite athletes*, women would continue to see gains sufficient to equal men such that in 1998 the fastest woman distance runner would be as fast as the fastest male distance runner was wrong.

The logic that was applied to these data was problematic for several reasons:

- The curves that were compared do not cover the same time periods because official records were kept for male marathons longer than for women (don't get us started...).

- The trend lines for women were derived from fewer data points than for the men (i.e., there are half as many triangles as circles).

308

CHAPTER 12   BUSINESS, RESEARCH, AND DESIGN RELATIONSHIPS: IT'S COMPLICATED

- Past events cannot be used to make predictions about future ones when you don't have equivalency (see above), and other factors are likely at play regarding the spread of the data. In this case, it is that socially women weren't afforded nearly the same opportunities as men (e.g., In the United States, Title IX legislation, which prevents schools from discriminating on the basis of gender, wasn't passed until 1972).

Indeed, if we continue with the logic that was used in that paper then, there is no reason to presume that men and women would reach their peak speeds in 1998. Put differently, why wouldn't women just continue with their upward trajectory of becoming faster and faster? By extending their model in this way (i.e., keep extending the dotted lines forward in time), one could predict that in the year:

- 2064: Women marathoners will be able to outrun male 100-meter sprinters

- 2271: Women marathoners will be able to outrun cheetahs

- 6419: Women marathoners will be able to break the sound barrier

- 103700: Women marathoners will be able to run fast enough to launch themselves into low-Earth orbit

So we need to be careful about how we interpret and use data. The good news is that you needn't worry about a pint of mint chip attracting sharks (we don't think), although you should still be careful out in open waters. It can be dangerous out there...and we can't rule out exactly what Ben and Jerry are up to.

309

CHAPTER 12   BUSINESS, RESEARCH, AND DESIGN RELATIONSHIPS: IT'S COMPLICATED

Data in Practice

Dr. Baker once took an executive team through a design thinking empathy workshop to address a problem that a customer was having around collecting attendance. She had video recordings of interviews with teachers describing their mornings, which were, to put it mildly, harried. They painted a picture of students coming in with questions, wanting to visit the teacher, or having to rush back out because they forgot something, while the teacher was:

- Trying to pull up their lessons for the day

- Deal with a slow network or computer issues

- Read and log parents' notes for prior children's absences

- Grade papers

- Read/listen to some training announcements from the districts

- Read a communication from the principal

- And, oh yes, they were taking attendance

It was stressful just listening to the descriptions! Viewing those videos definitely elicited a lot of empathy in the team (remember those mirror neurons?), which in turn evoked a strong desire to solve their problem.

Understanding the user context is key when trying to define and solve problems. You'll notice the basis of this step is, unsurprisingly, research. You need to gather data – videos, statistics, surveys, observations, etc. – to understand your user (and how they differ from you). One of the biggest mistakes we see in the Empathize step of design thinking is skipping the research and relying on anecdotal evidence (stories from the CEO's best friend's cousin's daughter who happens to be in an adjacent position to the one you're targeting) or nothing at all except your intuition. Gather your

310

CHAPTER 12 BUSINESS, RESEARCH, AND DESIGN RELATIONSHIPS: IT'S COMPLICATED

data, analyze it carefully, and then get ready to define your problem. And remember, anecdotes are not research.

EMPATHY IS A MISNOMER

Although this step is called "Empathize," it might be better termed "Understand" as you're not trying to feel the other person's pain or frustration – rather, you're trying to appreciate their motivations, circumstances, and challenges. If our arm is broken, we don't need the ER tech to "feel our pain" (we imagine that would keep them from being able to help us) – but we do need them to understand that we are in pain and to use their expertise to fix the problem.

Define: Problems Before Solutions

As human beings, we not only like solutions – we crave them.[5] Solutions are fun – they feel like accomplishments. They are neat and tidy. They make us feel smart. Unfortunately, problems are just the opposite. They feel unfinished. They're messy and complicated. They make us feel, well, dumb. As a result, we often find ourselves running straight toward "the answer" without taking a moment to consider what the question even was. Creating a solution without first defining the problem leads to a great deal of rework and failed projects. And yet we still do it. In large part because we don't know how to go about defining problems.

---

[5] Wong, P. T. (Ed.). (2013). *The human quest for meaning: Theories, research, and applications*. Routledge.

311

CHAPTER 12   BUSINESS, RESEARCH, AND DESIGN RELATIONSHIPS: IT'S COMPLICATED

Solution to a Problem or Solutions Looking for a Problem?

Let's take a look at some well-known products. As you think about each one, think – is this a solution to a problem or a solution looking for a problem?

1. Google Glass
2. Swiffer WetJet
3. Microsoft Clippy
4. Reebok Air Pump

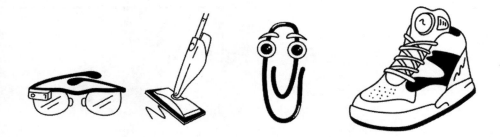

Got your answers? Great! Let's take a look at each one.

Google Glass

Google Glass was an experiment in wearable tech soft-launched in 2013. Its goal was to create a "ubiquitous computer." Failing to gain traction in the consumer market, sales were stopped in 2015. It pivoted to factory work seven years later), in 2020, where it also failed to get traction and was suspended in 2023.

Google Glass: Solution looking for a problem

Swiffer Wet Jet

Procter and Gamble had a problem – mopping chemical sales were not growing. They set their chemists to create a better soap, which ultimately failed. They then decided to better understand the consumer problem and set a team of design researchers to do in-home studies. During those studies, the researchers found something completely unsurprising – mopping sucks. You have to get a big bucket of soapy water, do the actual labor of mopping, and then clean the mop and the bucket. Anecdotally, while sharing a cup of coffee with a woman who had just finished mopping, one of the researchers spilled some coffee on the floor. Rather than haul the mop back out again, the woman simply grabbed a wet paper towel and cleaned up the mess. The Swiffer WetJet was born, creating a $500 million industry that changed how people clean floors.

Swiffer Wet Jet: Solution to a problem

Microsoft Clippy

Microsoft Clippy was an intelligent (and we use the term loosely here) office assistant launched in Microsoft Office 97 which was based on a misunderstanding of research that showed the same part of the brain responsible for mouse and keyboard interactions also triggered emotional responses. Ignoring focus group feedback that indicated it was "creepy," Clippy persisted in Microsoft Office until 2003 when it was finally removed. No, Clippy, we are not trying to write a letter.

Microsoft Clippy: Solution looking for a problem

Reebok Air Pump

Already a leader in the market, Reebok wanted to continue to grow and innovate. Through research, they discovered athletes were challenged by fits that were less than perfect – growing teens and adults alike needed something that would allow them to adjust their fit better than could be managed with laces. Inspired by the custom fit provided by ski boots, the

CHAPTER 12   BUSINESS, RESEARCH, AND DESIGN RELATIONSHIPS: IT'S COMPLICATED

Air Pump increased revenue by 18% and reduced injuries and improved performance for numerous athletes.

Reebok Air Pump: Solution to a problem

Defining Problems

So how do we approach a problem definition? At its simplest, a problem lies in the difference between the CURRENT state of a system and the DESIRED state of a system. The details of that difference may manifest in things like:

- Frustrations

- Inefficiencies

- Overlaps

- Low-skill, high-frequency tasks

But how do we uncover these things? And once we uncover them, how do we define them?

ROOT CAUSE

RB: Sometimes when I talk about problems with students, I'll offer the example that my foot hurts and ask "what should I do?"

JLR: So they ask lots of great questions about the pain and what you were doing before it hurt and…

RB: Nope. Usually they immediately start offering solutions: See a doctor! Take some ibuprofen! Get some ice! Stretch your foot!

JLR: And would any of that have worked?

RB: Nope, not even a little.

JLR: <smirking> So, why DID your foot hurt?

RB: I was standing on a Lego.

314

CHAPTER 12   BUSINESS, RESEARCH, AND DESIGN RELATIONSHIPS: IT'S COMPLICATED

It may be tempting to assume we "know" them (i.e., users) based on our own experiences, but personal experience is anecdotal at best and not a substitute for real observations. To find out what problems might exist in the space you are exploring you must do (cue ominous music) RESEARCH. Through interviews and observations, you must gather data on what your user is experiencing. Once you have your data, it's time to examine it. By breaking it apart, you can start to look for patterns and commonalities. Remember, even though we are looking for patterns, do not start to formulate a solution. Trying to leap to a solution before you've analyzed the data will cause you to ignore or overlook items in the data (or misinterpret the data). Things to identify during analysis include:

- **Context**: When/where/under what circumstances does the action happen? Are they outside or inside or both? Do the participants walk or run during this activity? Do they sit? Is their environment noisy or quiet? Are they trying to do other things at the same time? Is this activity confined to a short period of time – if so, what triggers it?

- **Action**: What are they trying to do? What steps do they take to do it?

- **Motivation**: Why do people do this activity? What leads them to it, and what benefit do they gain from it?

- **Definition of Success**: In a perfect world, how long does this activity take? What things should come out of this activity (what RESULT)? How accurate should it be?

- **Barrier**: What keeps them from getting the desired result? What is missing or in the way?

315

CHAPTER 12   BUSINESS, RESEARCH, AND DESIGN RELATIONSHIPS: IT'S COMPLICATED

- **Severity**: For the pain points within the activity, how bad are they? Is this a stubbed toe or a broken bone? Are we losing hundreds of dollars in contracts in a million-dollar business or tens of thousands of dollars in a million-dollar business?

- **Impact**: What impact does this pain have on the business? How does it affect the efficiency and effectiveness of the user? What does that translate into in business terms? Are we losing conversions, revenue, time, etc.?

- **Variables**: What changes or can be changed? What do we need to account for that might not hold steady from day-to-day? How much do they change (DELTA)?

- **Constants**: What can we count on to remain constant and steady?

Finally, it's time to create your problem. Remember, this is still not a solution! Take the items you've identified during your analysis, and use this template:

> When CONTEXT, ACTOR wants/needs to ACTION
> to RESULT because of MOTIVATION but cannot
> because BARRIER which causes IMPACT/SEVERITY

Here's an example:

> When plundering ships on the high seas
> [CONTEXT], pirates [ACTOR] want to attack
> only ships laden with rubber ducks [ACTION] to
> trade in Tortuga with the local gentry [RESULT]
> because there is currently a rubber duck shortage
> [MOTIVATION]. However, they cannot view the
> ships' contents without boarding them [BARRIER]

316

CHAPTER 12   BUSINESS, RESEARCH, AND DESIGN RELATIONSHIPS: IT'S COMPLICATED

which causes them to <u>board a lot of ships that have</u> <u>less desirable items such as gold and jewels, which</u> <u>in turn keeps them from turning a profit in the</u> <u>rubber duck market and costs extra time and pirate-</u> <u>power</u> [IMPACT/SEVERITY].

Notice, we still haven't come up with a solution yet (feels weird, right?). However, we've identified and **quantified** the problem sufficiently that when we create a solution, we should immediately be able to see whether it will be successful. The problem definition is key to reducing time and costs and is the critical first step to effective product development. Without problems, we end up creating solutions in search of a problem. And that's a problem.

IT'S NOT ABOUT YOU

Notice that problem definition here is done from the perspective of the user. The most common challenge I see when teaching this technique is to start viewing problems from the company's perspective. In one workshop, I had a group that defined the problem as falling sales numbers. This led to some rather creatively awful solutions of questionable ethics, not the least of which was a forcibly applied temporary tattoo advertisement intended to raise awareness of the product. It was both hilarious and disturbing. To be clear: Lower sales numbers is a problem. But it is the company's problem NOT the user's problem. And, while it is important to solve the company's problem so you can continue to get paid, you will do that by first solving the customer's (a.k.a., the user's) problem. By focusing on the user problem, you will be able to come up with good solutions that will solve the company's sales issues too. And that preferably do not involve forcing temporary tattoos on anyone.

317

CHAPTER 12   BUSINESS, RESEARCH, AND DESIGN RELATIONSHIPS: IT'S COMPLICATED

The next step is to define what success looks like (still no solutioning – sorry, not sorry!)

Defining Success Criteria

Why should you take time to define success before you solution? It's important to understand how you will measure success so you know whether your solution works and to agree, as a team, on what success looks like. That is, you need to both measure success and define it.

- **Defining Success**: By defining success criteria at the user problem level, you know what compromises you can make – and which ones you can't. It prevents scope creep and lets you know when you're done. And most importantly, it ensures that everyone is on the same page for what the solution must accomplish. If the design team thinks that success means the interface includes animations that engage the user, while the development team thinks success means that load times are shorter, and the product team thinks success means increased sales, there's going to be trouble. Agreeing what success looks like prevents these disconnects and ensures better outcomes.

- **Measuring Success**: If you do not define success criteria early, you risk being unable to measure the results of the solution because the way to measure it was not determined early enough to include in the system. Consider a solution which is intended to improve how quickly someone is able to complete a task. If you have not included analytics to measure how long it takes to complete the task and taken measurements of how long it takes today, you will end up with a solution that may or may not have fixed the problem.

How do you define and measure success?Success can be

- **Removal of the Barrier**. For example, if the barrier was that the user couldn't find the start of a workflow, removal of the barrier might mean that the user can find the start of the workflow without help. Measuring the removal of a barrier usually entails testing whether the user can now do the thing they could not do before as well as reduction in support calls.

- **Bypass of the Barrier**: For example, if the barrier was that the user couldn't find the start of a workflow, bypass of the barrier might mean that the user does not need to access the workflow at all. Measuring the

CHAPTER 12   BUSINESS, RESEARCH, AND DESIGN RELATIONSHIPS: IT'S COMPLICATED

bypass of a barrier usually entails testing the efficacy of the alternative (does the new workflow take less time than the old one, can it be found easily, or if the solution is to remove the workflow, is the result of the workflow still available or have all the items it supported been removed) as well as reduction in support calls.

- **Reduction of Impact/Severity**: For example, if the barrier was that the user couldn't find the start of a workflow, reduction of impact/severity might mean that the result of the workflow is no longer needed or that it is only needed once a year. Measuring reduction of impact/severity usually entails testing relative measures, such as reduction of time on task, reduction of complexity, improvements in findability, and reduction in errors and assists, as well as reduction in support calls.

Recall our duck-seeking pirates. Success for them might mean

- Being able to see the contents of any ship without boarding it (removal of the barrier)

- Having information about the contents of any ship prior to approaching it (bypass of the barrier)

- The ability to extract loot from ships that takes less time and effort than currently (reduction of impact/severity)

Ideate: Finally – Solutions!

Once you have your problem and your definition of success, it is FINALLY time to start solutioning. For most people, this is their favorite part. In fact,

CHAPTER 12   BUSINESS, RESEARCH, AND DESIGN RELATIONSHIPS: IT'S COMPLICATED

this is such a popular part that many agencies and companies skip straight to this part, eschewing data collection, analysis, and problem definition entirely. Say it with us: "Don't be Clippy." Successful solutioning relies on a solid foundation of research and problem definition.

There are hundreds of ways to go about solutioning (brainstorming, SCAMPER, crazy 8's, and Worst Possible Idea are just a few). Each of these techniques has in common that you will come up with as many ideas as possible, many of which will be terrible, all of which could solve the problem and fulfill the stated success criteria. This is your chance to be creative and a little whacky – do not be afraid of terrible, ridiculous solutions! While they might seem like a waste of time, often the most outrageous solutions will spark ideas on how to elegantly address the issue. Draw your solutions out as cartoon storyboards or single panels to help others visualize what you were thinking. Don't worry about whether or not your drawings are realistic – stick figures and diagrams work great. Remember, you want to generate a volume of ideas that you can then sift through to find the gold (or, better yet, rubber ducks).

Once you have defined as many solutions as possible, evaluate them as a team. List their pros and cons, compare them against the problem to be solved and success criteria, and choose the top five. As a team, flesh out each of those top five, in as much detail as possible. How much might this solution cost in time and development? How well does this solution solve the problem? Does this solution also solve other problems? What is the longevity of this solution? Will it scale? What will it take to maintain this solution? After answering as many questions as you can about the solutions, narrow them down to one that you can afford to build and maintain and that will solve the user's problem. This is the one you will want to prototype.

CHAPTER 12 BUSINESS, RESEARCH, AND DESIGN RELATIONSHIPS: IT'S COMPLICATED

Prototype: Try Before You Buy

There's a famous team building exercise in which teams are asked to build the tallest freestanding structure using 20 sticks of spaghetti, a yard of tape, a yard of string, and one marshmallow in 18 minutes. The marshmallow has to be on top. Developed by Tom Wujec, the exercise delves into how teams work together and specifically the value of prototyping. Successful teams will generally experiment a lot – that is, they'll jump in and build a number of prototypes to test out ideas on balance and structure early on to inform their final structure. This exercise is what often drives the "creative" exercise in design thinking of constructing things with pipe cleaners and other arts and crafts materials. The physicality of the exercise is not what is important however – rather, it's the creation of the idea. Sketching out potential screens would work as well (and involves a lot less glue). The key is to create a prototype you can test.

Using experimentation to refine and test ideas in a low-cost environment helps ensure you ensure your failures are affordable and low-impact. And you will have failures. No great idea has ever been successful on its first run. You have to try it, find out all the stupid things you forgot to take into account, and try again before you are successful. Most prototypes will not involve comestibles like spaghetti and marshmallows but rather be sketched out on paper or drawn up in Figma (or the design software of your choice). By taking the time to flesh out the primary pieces of your proposed solution, you can show them to prospective users and get early feedback. Prototypes are cheap, easy ways to test ideas and make sure they'll work.

Test: Measure Twice, Cut Once

You may remember that way back in Chapter 1 we talked about how a lot of design research doesn't provide results that are of use. Part of this is due to the small sample sizes that often get used. Now, we're not espousing

CHAPTER 12 BUSINESS, RESEARCH, AND DESIGN RELATIONSHIPS: IT'S COMPLICATED

that all user testing must meet the rigorous standards set forth by academic journals – not even close. What we are espousing (because it is fun to use that word twice in as many sentences) is that you need to have confidence that what you are seeing in the data is actually due to the factors that you think and not just chance variation. This is why conceptually it is important to know what significance testing, which is what academic researchers rely on, is trying to do.

At its core, significance testing is used to allow researchers to understand how likely it is that a result could just happen on its own. If we think about drug testing, a pharmaceutical lab would compare the percentage of people who get better or have a reduction of symptoms and the percentage who received a placebo (thus who just got better on their own). The significance test then provides insight into whether the difference between those two percentages has *the potential* to be meaningful or not. There is never any certainty because we are talking about complex probability functions that go from negative infinity to infinity (stay with us), but there can be confidence that what we are doing has a good chance of working (or not).

The same holds true when you test your designs at a conceptual level: you want to know that changes to flow, navigation, etc. are due to what you've done and not just circumstance. The way to increase your level of confidence *without conducting significant testing* is to test on as many individuals as is feasible.

Which means you need to ask yourself: "Am I able to be confident after only testing or talking to five people?"

Yet, this is what many in design will do: ask or test five people and then stop. But, what if there is something about those five individuals – perhaps they all took the same highway to the testing facility and saw an accident or were stuck in the same snarl of traffic or randomly all just reacted the same way, etc.?

And where did this number five come from anyway? We can tell you that it is NOT accepted as some magic number or standard practice by

323

CHAPTER 12   BUSINESS, RESEARCH, AND DESIGN RELATIONSHIPS: IT'S COMPLICATED

anyone outside of design who conducts research and, honestly, should not be accepted within design, as the following study should make clear.

A comparison of the results of user testing groups of various sizes was conducted in order to see how many usability errors would be identified.[6] Starting with a population of 60 user testers, the researcher found that when groups only consisted of a random assortment of five individuals the groups failed to identify *almost half of the issues,* but when the size of the testers was increased to 20, 95% of the errors were successfully identified.

Are you confident that if you were to test more individuals that the results would still be the same as you saw with those five?

Furthermore, it should be noted that the original work that proposed this "you only need five users" idea did so in the context of how many people you needed to identify errors, which is similar to the study referenced above. What was not suggested (at least initially) was to only use five user testers to determine sentiment (i.e., how do people feel about the designs) or more complex behaviors.[7] Think about it – would you feel safe knowing that a drug was only tested on five people? A new airplane safety mechanism was only tested on five flights? Even that a new flavor of chewing gum (mmmm...artichoke!) was found to be preferred by four out of five individuals (compared to black licorice...yuck)?

This is why before any testing is carried out (again – on more than five people), you need to determine the threshold that you are comfortable with when interpreting the results. As a useful starting point, you could just go with chance. So for something with two possible outcomes, like tossing a coin, you know that chance is 50%. Thus, you may feel that the minimum

---

[6] Faulkner, L. (2003). Beyond the five-user assumption: Benefits of increased sample sizes in usability testing. *Behavior Research Methods, Instruments, & Computers, 35,* 379–383.

[7] Turner, C. W., Lewis, J. R., & Nielsen, J. (2006). Determining usability test sample size. *International encyclopedia of ergonomics and human factors, 3*(2), 3084–3088.

CHAPTER 12   BUSINESS, RESEARCH, AND DESIGN RELATIONSHIPS: IT'S COMPLICATED

threshold you are comfortable with is at least 75%, meaning that three-fourths of user testers should feel a particular way or have a particular outcome (e.g., faster navigation time) in order for you to feel confident in the results.

After you've decided on your threshold, then you need to decide on the number (i.e., "N") you will use to gain this confidence. There are certainly statistical techniques that could be used to do this, but that aside, if we are keeping this simple, then we also need to keep it intellectually honest. Here is what "keeping it honest" looks like:

You've decided in advance (or *a priori* as the scientists like to say) that you will feel confident in your results if you meet or exceed a threshold of 75%, and you'll test your new designs with at least ten users. You are able to get 11 users to go through an unmoderated session, and you find that of the 11, 8 yield the results you're looking for. Great, right?

Nope. You set your threshold at 75% and 8/11 is 72.7%. You did not meet the threshold that *you set* for confidence.

Now, when we say keep it "intellectually honest," what we mean is that your next step shouldn't be to go out and get just one more user tester and hope that they give you the result you want (which, if they performed the way you hope, would raise the percentage to 9/12 or 75%). To do this would be a knockoff version of what researchers have started to self-police and warn against, which is an unethical practice known as *p*-hacking[8] (the "p" stands for "probability testing").

Does this mean all is lost? Not even close. You should talk to the three dissenters and determine why their responses looked different. Look back at the data and see what patterns there may be. Who knows there could be something super interesting lying in the absence of the result (see the "lizards and beetles" story from Chapter 1).

---

[8] Head, M. L., Holman, L., Lanfear, R., Kahn, A. T., & Jennions, M. D. (2015). The extent and consequences of p-hacking in science. *PLoS biology*, *13*(3), e1002106.

CHAPTER 12   BUSINESS, RESEARCH, AND DESIGN RELATIONSHIPS: IT'S COMPLICATED

Design Thinking: Are All the Sticky Notes Worth It?

Followed carefully and fully – doing research to understand the user context, closely defining the problem they are experiencing and what success looks like, solutioning creatively, creating prototypes to test and then testing them (and being willing to throw away failures) – design thinking can be very effective and help companies get "unstuck." HOWEVER, design thinking is not a substitute for research and is not a quick fix to your design problems. Much like design systems, design thinking is touted as a magic pill that will fix your customer woes, but like design systems, it requires a careful, methodical, fact-based approach to be effective. It takes time and investment. As such, you should not try to apply design thinking to every design problem. Design thinking is effective for complex problems because it lets you tease insights from the data, closely define the problem you're trying to solve, and be creative within challenging constraints. For simple problems, it is likely not worth the time and investment.

Timing Isn't Everything – It's the Only Thing

In real estate, the top three drivers for a good sale are "location, location, and...location." For research, those top three are timing, timing, and . . . timing. Asking to do discovery research when the product is about to ship and preparing to do A/B testing on something that is still a prototype are pointless exercises because the time is not right for the technique – and believe it or not, these are common asks by product managers and executives to research teams. It's important to understand what questions you want to answer based on WHEN you are in the product life cycle, because not all time segments are created equal.

326

CHAPTER 12   BUSINESS, RESEARCH, AND DESIGN RELATIONSHIPS: IT'S COMPLICATED

What Should I Ask When?

Asking the right question at the right time is key. In the following diagram, you can see what type of questions you might want to answer at different stages.[9]

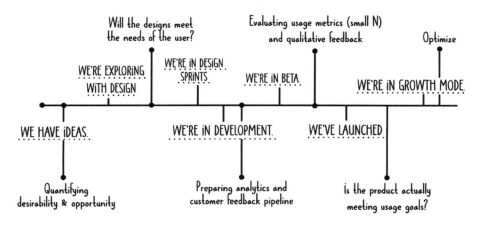

For example, at the beginning of a product cycle, you've got ideas. This is a great time for discovery work: Do these ideas resonate with others (desirability)? Are there other products like this in the market (opportunity)? As you can see, it would be silly to ask these questions once you've launched the product! Knowing what you need to find out will in turn inform the type of research technique you can use to get those results. The following diagram outlines some of the techniques available to you at different time segments:

---

[9] Credit for this way of looking at research goes to Ben Doctor.

CHAPTER 12   BUSINESS, RESEARCH, AND DESIGN RELATIONSHIPS: IT'S COMPLICATED

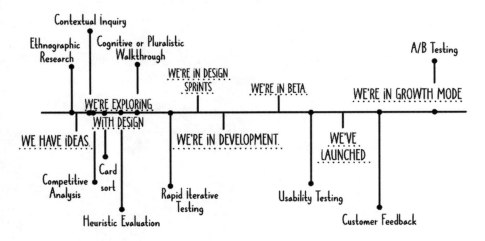

Let's take a moment to quickly define some of these techniques and their goals:

Type	Description	Goal
Cognitive walkthrough	Conducted by internal evaluators working through hypothetical user tasks and adopting hypothetical user mindsets. The context/intent of the product is taken into account.	Provides high-level feedback on overall navigation, content of the product.
Rapid iterative testing	A designer, researcher, and user will conduct discrete usability sessions together. And between sessions, designers will rapidly incorporate feedback from previous sessions.	Provide fast feedback on discrete features and workflows.

(continued)

CHAPTER 12 BUSINESS, RESEARCH, AND DESIGN RELATIONSHIPS: IT'S COMPLICATED

Type	Description	Goal
Usability testing	A test conducted to examine whether users can accomplish a given set of tasks with the product.	Provide detailed, fine-grain feedback on the usability of a set of tasks.
A/B testing	Compares two versions of the same interface for the same task and determines which is easier/more desirable. Requires significant number of participants to be valid.	Understand how specific design iterations relate to user actions.
Heuristic evaluation	A method of evaluating interfaces using established heuristics. The context/ intent of the product does not affect the heuristic criteria that can be used universally.	Provide targeted feedback of areas that violate well-known usability practices.
Competitive analysis	Examines how we solve user problems as compared to alternative competitor products. Looks at interactive elements, IA, and/or visual cues.	Provide insight into comparatively desirable and undesirable interactions.
Ethnographic studies	Onsite observations of user' work and environment to understand the context and motivation around the work.	Provide insight into the work habits and environment of the user, uncovering areas of friction or opportunities that may not be self-reported.

(continued)

329

CHAPTER 12   BUSINESS, RESEARCH, AND DESIGN RELATIONSHIPS: IT'S COMPLICATED

Type	Description	Goal
Interviews	In-person or remote interviews with representative user to understand more about specific areas of expertise, product, or process.	Provide clarification or understanding of areas that are the expertise of the user around how they use the product, what motivates them to use the product, and what pain points they experience.
Contextual inquiry	In-person or remote walkthroughs of a prototype with a representative user to test ideas and workflows at an early stage.	Provide insight into proposed workflows to determine applicability and desirability as well as match to context.

CHAPTER 12   BUSINESS, RESEARCH, AND DESIGN RELATIONSHIPS: IT'S COMPLICATED

Timely Research Is Timely

A lot of companies will eschew research because they feel it takes too much time. This reaction often stems from poor experiences in the past in which research was requested at the last minute, at the wrong time, or (possibly worst of all) when the only thing desired was confirmation. When asking for research, ensure you can answer the question "What will I do differently because of this?" before you continue. You have to be prepared for getting results that you do not like – for example, that no one is interested in the new feature, that the new workflow is harder to use, etc. You have to get results in a time frame during which you can react to

CHAPTER 12   BUSINESS, RESEARCH, AND DESIGN RELATIONSHIPS: IT'S COMPLICATED

the new information – for example, stopping development of a feature, redesigning the workflow, and so on.

That said, researchers must understand that precision is not as important as timeliness. Being able to gather and provide results quickly is key to informing product design and development, with sooner always being better than later. Ensure you establish your confidence threshold prior to starting and stick to it – don't let perfect be the enemy of good enough.

Recap

- Design thinking is a solution-focused framework with five stages: Empathize, Define, Ideate, Prototype, and Test. Doing design thinking correctly is time- and resource-intensive but is very effective for complex problems. It is not a cure-all (nor does it need to be applied to every problem)

- Having data is not the same thing as using data effectively. Correlation does not equal causation. Ben and Jerry are probably not super villains.

- Problem definition is key in reducing time and costs and is the critical first step to effective product development and using research optimally.

- When testing, you need to determine your confidence threshold. What level of uncertainty are you comfortable with?

CHAPTER 12   BUSINESS, RESEARCH, AND DESIGN RELATIONSHIPS: IT'S COMPLICATED

- Depending where you are in the product development life cycle informs what kind of questions you are able to answer with research and the techniques you can use to answer those questions.

- Data is only useful when it's useful – that is, you have to have the right data to answer the right question at the right time. Having data isn't helpful if you are not at a point at which you can apply it.

Before You Go...

A company Dr. Baker was working for decided to engage a high-end design firm to help reimagine some of their enterprise products, making them more modern and appealing – sales had been slow, so it was assumed that

CHAPTER 12 BUSINESS, RESEARCH, AND DESIGN RELATIONSHIPS: IT'S COMPLICATED

the issue was the interface was outdated. As part of the discovery process, the team did a number of site visits to gather data on the users to better understand their day-to-day challenges. Walking into the office, the first thing she noticed was that each of the soul-sucking grey cubicles was stocked with huge, three-ring binders, bursting at the seams with papers, yellow sticky notes marking multiple saved sections throughout. As we settled in for the interview, she could see that the technology they were using was old – at least five years out of date. The interview she remembers best was a gentleman in his fifties. He was not at all interested in the team making any changes to the system he used, and as the interview went on, he became increasingly annoyed. As the interviewer asked what features he would want in a mobile version of the product, he finally had had enough. He picked up his flip phone and waved it under the interviewer's nose.

"See this?" He asked. "My wife made me get this so she can get a hold of me. That is the ONLY thing I do on my phone. I don't need to do anything else, and moreover, I don't WANT to do anything else. If I can do work from my phone, people will start expecting me to work nights and weekends! There's nothing going on here that can't wait until I get back in the office. I need a mobile interface about as much as a NASCAR driver needs a right turn signal! I don't need any of this fancy stuff – I just want this system to do what I want, when I want it, every time."

The other interviews went similarly, although with less colorful analogies. After the team got the data, they went back to analyze and ideate. They came back with solutions that were beautiful, creative, and completely at odds with running on outdated systems with throughput issues. They had responsive versions of the interface and serious graphics-intensive views that would need a high-end gaming computer to render. After spending millions of dollars, ultimately, none of those designs were implemented. The designers and the company had failed to effectively understand the user, their context, and the problem to be solved. Although they did data collection, they failed to use the data that didn't agree with their original assumption/hypothesis. And as a result, they ended up building a right turn signal for a NASCAR driver.

334

CHAPTER 13

The AI Elephant in the Room

> *You know what the biggest problem with pushing all-things-AI is? Wrong direction. I want AI to do my laundry and dishes so that I can do art and writing, not for AI to do my art and writing so that I can do my laundry and dishes.*
>
> —Joanna Maciejewska, March 29, 2024 `https://x.com/ AuthorJMac/status/1773679197631701238?lang=en`

RB: I'm constantly amazed at how people either love or hate generative AI. There seems to be no middle ground – it's either the death of society as we know it or the birth of a new age of enlightenment.

JLR: There's definitely a lot of emotion and fear rolled into the reactions we see today. But that's hardly new. Think of Socrates and his objections to the written word. He was convinced that it would make people lazy thinkers because they no longer had to remember things.

RB: Great example! Advances in technologies often come with a lot of fear and trepidation – cars, industrial machinery, television, personal computers, mobile phones – each came with its own set of doomsayers, convinced that the technology would steal our jobs and our ability to think and undermine our social structure.

© Jerome L. Rekart and Rebecca Baker 2025
J. L. Rekart and R. Baker, *Designing for Human Intelligence in an Artificial Intelligence World*,
https://doi.org/10.1007/979-8-8688-1418-1_13

CHAPTER 13   THE AI ELEPHANT IN THE ROOM

JLR: Especially when we seem to do a fine job of undermining our ability to think and advance without the help of technology. Current readership aside ;)

RB: Well said.

Different Timetables: Moore's law vs. Evolutionary Change

"Kids today are just born knowing how to use these things" was a statement overheard by Dr. Rekart as he observed one harried parent talking to another as she looked at her child using a handheld device. This sentiment, which is bandied about in many casual circles, suggests that not only will technology somehow change *us* biologically but that the content of this book as it applies to human nature won't be viable for too much longer either! Luckily (or not), there is a next to zero chance that either of those two situations will come to bear. Why not? Well, we've already covered the fact that there aren't actual biological differences between different generations (see Chapter 7), but what we didn't address was why this is the case and why even though the technological world around us may change dramatically and drastically, we won't grow (within multiple lifetimes, at the least, if ever) to accommodate those changes as a species.

A straightforward way to show why the arrow of causality (i.e., humans affect technology) only points in one direction is to examine two processes that govern change within each, beginning with modern computer technology.

In 1965, Intel cofounder Gordon Moore predicted that the number of transistors contained within an integrated circuit would double every ten years (which he later revised to a doubling every two years).[1] This observation and prediction, which has come to be known by the eponym

---

[1] Moore, G. E. (1965). Cramming more components onto integrated circuits. *Electronics, 38,* 114–117.

CHAPTER 13   THE AI ELEPHANT IN THE ROOM

"Moore's Law," captures the ever-increasing computational power possible with artificial thinking machines (i.e., computers). To illustrate this "law" practically, one only needs to consider that in 1985 the CRAY-2 was the most powerful computer on the planet and was decidedly not a consumer-level product that could be considered "personal." It took up about 16 square feet and weighed over 2 tons! In less than 40 years, however, the Apple iPhone 12, which could be found in the lockers and pockets of teens everywhere across the planet, was 5,000 times more powerful[2] but weighed less than 6 ounces. Think about all of the innovations, models, versions, etc. that were developed between the time of the CRAY-2 and the iPhone 12. Think about the size that computers take, the types of operations they can fulfill, and the speed with which they function (who among us remembers the hours – and dozens of CDs or floppy disks – that updates to new programs or operating systems would entail). The pace of advancement has been rapid and pronounced.

Now, let's contrast the time frame and number of iterations between technological advancements and what has occurred evolutionarily between the organic equivalent of a supercomputer (i.e., the human brain) and its closest extant analog, a chimpanzee. The difference between chimpanzees and humans, which at a genetic level is between 1 and 3%,[3] is measured by the time that took place between now and when the two species diverged from a common genetic ancestor. According to modern molecular dating techniques, this event took place between 6.5 and 7.5 *million* years ago.[4] Now, this is a tricky comparison to make given that

---

[2] https://blog.adobe.com/en/publish/2022/11/08/fast-forward-comparing-1980s-supercomputer-to-modern-smartphone

[3] The range is a byproduct of whether you're looking at coding or noncoding regions of nuclear DNA.

[4] Suntsova, M. V., & Buzdin, A. A. (2020). Differences between human and chimpanzee genomes and their implications in gene expression, protein functions and biochemical properties of the two species. *BMC genomics, 21*(Suppl 7), 535.

CHAPTER 13   THE AI ELEPHANT IN THE ROOM

there were closer analogs (i.e., Neanderthals and Denisovans), but we can only infer how smart they were. Regardless, the point is that evolution – the pace of our advancement – is on a completely different timescale.

This means that even if technology were able to have an impact biologically on our systems (which would entail some sort of selective advantage), the process would operate on a timescale such that the technology that would engender the change would be obsolete by the time that humans evolved sufficiently to take advantage of it.

Self-Correcting...or Not

At the time of this writing, over almost a quarter of a million games of major league baseball have been played in the United States.[5] Despite the sheer number of games played, something never before seen happened over the summer of 2024. This feat was that the same player, a catcher by the name of Danny Jansen, started the game for one team (the Toronto Blue Jays) and finished the game on the field for the opposition (the Boston Red Sox).[6] Now, this is possible because the game in question was actually played on *two* different days, which happened to be separated by a span of 1,464 hours (June 26 and August 26). The fact that the same game was spread across multiple days is not *that* unusual as games that have to be suspended due to rain or other weather conditions are routinely reconvened at a later day and time (if a game doesn't make it to the middle of the fifth inning before being suspended, it isn't considered "official" and must be resumed when possible). No, what was unusual about this

---

[5] https://www.baseball-reference.com/leagues/index.shtml
[6] https://www.nytimes.com/athletic/5764512/2024/09/13/danny-jansen-game-dodgers-diamondbacks-first-inning/?source=user_shared_article

CHAPTER 13   THE AI ELEPHANT IN THE ROOM

particular situation was that at no point in time had a player ever been traded from one of those teams to another and thus have the ability to contribute to both teams' chances of winning!

Now, other than being a neat piece of baseball trivia, what is important for our focus is not the game itself but the fact that modern computers, which track game statistics, couldn't accommodate the fact that the same player could "simultaneously" be a member of both competing squads. The computers used by one site (Baseball Reference) couldn't understand something that could be explained and grokked by you in the time it took you to read the previous paragraph. This idea – that novelty is something that we can quickly accommodate – is a hallmark of our original intelligence and is important to keep in your (original) mind as we dive into our discussion in the remainder of this chapter.

Bias In, Bias Out

Users and developers of technology have long known the adage "garbage in, garbage out." Though this pithy statement certainly reminds us to be mindful of what we are using as the foundation for our creations, it doesn't truly capture what happens when we don't fully think through what we are asking of our technology and whether *we* have set it up for success. Consider recent advances that have been made using artificial intelligence to diagnose melanoma (a serious, often deadly, form of skin cancer). Initial models were touted as being as good – if not better – than trained human dermatologists at diagnosing cancerous lesions from images.[7] Though this seems like a fantastic advancement (and it is certainly a promising *future* use case for AI), it was later shown that those models were trained on a data set that primarily contained examples of lesions from white

---

[7] Brinker, T. J., Hekler, A., Enk, A. H., Berking, C., Haferkamp, S., Hauschild, A., ... & Utikal, J. S. (2019). Deep neural networks are superior to dermatologists in melanoma image classification. *European Journal of Cancer*, *119*, 11–17.

339

CHAPTER 13   THE AI ELEPHANT IN THE ROOM

patients, for whom melanoma can present differently than it does with black patients.[8] The result of this bias in the underlying training data was a halving of the diagnostic accuracy when examining lesions from patients of color[9] – an untenable oversight.

Though presumably most of our design situations will not – luckily – be a matter of life or death, it is important to ask ourselves about the possible biases that may underlie an AI system that we are using. As with the previous example, was it trained in such a way that its use wouldn't be equally adept? This should definitely be done with AI, but certainly isn't restricted to only that domain. Psychological researchers will often ask whether the underlying data for a finding or set of results are "WEIRD." This tongue-in-cheek acronym stands for "Western, Educated, Industrialized, Rich, and Democratic" and is a useful way to ask whether culture, privilege, or other factors may affect the interpretation of behavioral data.[10].

Personification of All the Things

The first and third of Asimov's three laws of robotics might as easily be considered rules to live by for human beings, pointing out a double standard we have for what artificial intelligence might become in the distant future.

1.   *A robot may not injure a human being or, through inaction, allow a human being to come to harm.*

---

[8] Norori, N., Hu, Q., Aellen, F. M., Faraci, F. D., & Tzovara, A. (2021). Addressing bias in big data and AI for health care: A call for open science. *Patterns*, *2*(10).

[9] Kamulegeya, L., Bwanika, J., Okello, M., Rusoke, D., Nassiwa, F., Lubega, W., ... & Börve, A. (2023). Using artificial intelligence on dermatology conditions in Uganda: A case for diversity in training data sets for machine learning. *African Health Sciences*, *23*(2), 753–763.

[10] https://www.apa.org/monitor/2010/05/weird

CHAPTER 13   THE AI ELEPHANT IN THE ROOM

2. *A robot must obey orders given it by human beings except where such orders would conflict with the First Law.*

3. *A robot must protect its own existence as long as such protection does not conflict with the First or Second Law.*

—Isaac Asimov, Three Laws of Robotics, *Runaround*, 1942

Science fiction authors have long used the idea of artificial intelligence and the implications as material for stories and films. The idea of something built rather than born provides a perfect platform for exploring the origins of morality, ethics, societal responsibility, and other equally esoteric philosophical topics that are not easily quantified. How can something that springs into existence without a literal lifetime of lived experience be expected to behave? What is the origin of self-sacrifice? The AI in these stories can be cold and logical, like the agents in *The Matrix*, SkyNet in *The Terminator*, or HAL from *2001*. Or they can be sympathetic like Frankenstein, Sonny from *I, Robot*, Number Five in *Short Circuit*, or *Wall-E*. But, no matter whether the AI/robot/creation is good or evil, one thing they all have in common is that they are fictional. The possibility of a ruthless agent from *The Matrix* or an adorable Number Five attaining consciousness while seeking to be understood is definitely not a current reality. However, we persist in interacting with generative AI in ways that suggest we view them as, well, human.

Personification is our tendency to assign human motivations and characteristics to nonhuman entities. We frequently do this with pets and objects in our everyday lives, assigning them motivations and emotions that are beyond their capacity. We say an animal that rips up a couch or pees on your shoes is "doing it to punish you." Your word processing application isn't working properly because "it's having a bad day." Your car won't start because it's "being difficult." The truth is, none of these statements are true. Computers, cars, and

341

CHAPTER 13   THE AI ELEPHANT IN THE ROOM

pets don't have the same emotional and motivational capacity and range as a human being. However, it comforts us to assign those types of characteristics to them to (1) provide an easy context for the observed behavior and (2) mitigate the amount of responsibility that we, ourselves, bear for the behavior. The same tendency holds true for AI. We most often go awry with AI when we assume it has motivations or a grasp of esoteric concepts like "truth." The examples of AI gone-wrong in this book – the diverse Nazi pictures, the virtual priest offering to baptize your child in Gatorade, the poor word choice in advertisements – stem from the system behaving as it was designed but not as a human. And we are surprised and appalled that it did exactly what we programmed it to do without regard for our motivations or context.

THAT'S MY PIE!

As an owner of two adorable fur babies, I deeply understand the desire to personify their actions. We have one dog that is very food-focused – to the point that I found him (all 75 lbs of orange and white fur) in the middle of the dining room table snuffling for crumbs. Unsurprisingly, someone will occasionally be incautious and leave some tasty treat out on the counter unsupervised – most recently, our oldest left a pumpkin pie out on the counter then walked into another room to do something. Ever ready to take advantage of a food-based opportunity, the dog promptly got up on the counter and ate half the pie before the (now outraged) child returned to rescue his pie and shoo the dog out of the kitchen. Said child was furious (pumpkin pie being his favorite) and accused the dog of "stealing my pie to get back at me for not petting him more." To be fair, the child in question knows that's not true, but in the moment, he reverted to personification to provide a context for the behavior in which he was not wholly culpable.

CHAPTER 13 THE AI ELEPHANT IN THE ROOM

Soft Skills Are Hard

I have people skills; I am good at dealing with people. Can't you understand that? What the hell is wrong with you people?

—Tom Smykowski, Office Space

People skills, also referred to as soft or durable skills, comprise a set of skills like communication and critical thinking. First defined by the US army[11] to distinguish between interpersonal skills and skills for using machinery or weapons, soft skills have become increasingly important in the workforce, reflecting a continuing move from industrial to information workforce jobs.

Between 1980 and 2012, social skill-intensive occupations grew by nearly 12 percentage points as a share of all U.S. jobs. Wages also grew more rapidly for social skill-intensive occupations than for other occupations over this period. In contrast, both employment and wages grew more slowly for occupations with high math but low social skill requirements, including many STEM jobs. Directly comparing the returns to social skills in the NLSY 1979 and 1997 surveys, I find that social skills are a significantly more important predictor of full-time employment and wages in the more recent cohort.

——Deming, D.

"The Value of Soft Skills in the Labor Market"

The Reporter, NBER, Jan 5, 2018

https://www.nber.org/reporter/2017number4/
value-soft-skills-labor-market

---

[11] https://www.britannica.com/money/soft-skills

CHAPTER 13   THE AI ELEPHANT IN THE ROOM

This noted growth has only continued as the availability of virtual work has continued to expand as noted in the 2020 World Economic Forum Report.[12] Being able to demonstrate solid durable skills are increasingly being expected by employers[13] and have been linked to successful advancement in technological fields (typified by their reliance on hard skills).[14] However, this is one of those interesting areas in which we all nod sagely about the importance of a thing (soft skills) but fail to see that we perhaps lack that thing or that we need it in our current situation.

As an example, a study done in Vietnam[15] highlighted that students did not see the need for soft skills for success in academics while recognizing it as important for career success. Perhaps the most apt example is that of the Talented Jerk – an individual who provides great business value to the organization through their application of hard skills but is toxic to the team and culture due to their lack of emotional intelligence and durable skills. A common management issue,[16] the Talented Jerk is a prime example of why communication skills are important in any environment that involves...well...communication, and why the most common reason that they are difficult to teach is that most people lacking these skills have no idea that they are lacking them. Put another way, the villain in the story very rarely thinks of themselves as the villain.

---

[12] https://www.weforum.org/publications/the-future-of-jobs-report-2020/in-full/

[13] https://businessnamegenerator.com/learning-hub/soft-skills-in-the-workplace/

[14] https://pmc.ncbi.nlm.nih.gov/articles/PMC10428053/

[15] https://www.researchgate.net/publication/379672636_The_Importance_of_Soft_Skills_for_Academic_Performance_and_Career_Development-From_the_Perspective_of_University_Students

[16] https://www.forbes.com/councils/forbescoachescouncil/2024/07/30/managing-workplace-bullies-how-to-lead-a-brilliant-jerk/

CHAPTER 13   THE AI ELEPHANT IN THE ROOM

At one point in her career, Dr. Baker was collaborating with a colleague on a paper for a conference. The data the paper was based on had been gathered already – all that was left was the analysis and write up. Her colleague went to his manager to let her know he was involved in the paper and would be presenting at the conference. The manager promptly threw a fit, angry that he had not asked permission to work on a paper. She then forbade the colleague from working on the paper and from going to the conference. The paper, which would have benefited the company's reputation in the industry, remained unwritten and unpresented because of the manager's lack of foresight, empathy, and humility – all important soft skills.

And the need for soft skills is only increasing. As more businesses moved to a virtual presence during the pandemic in 2020, the advantages and challenges of virtual communication became more apparent. A lack (think email/messaging) or reduction (think video calls) of physical cues, often taken for granted in face-to-face (F2F) communications, can result

345

CHAPTER 13   THE AI ELEPHANT IN THE ROOM

in increased misunderstandings and missed information.[17] Understanding how to interact in a virtual-enabled environment is arguably another (albeit related) type of communication skill. A quick search on virtual communication training programs brings up dozens of hits that will help you write emails, run meetings, and more.

AI and (Not So) Soft Skills

Although it seems obvious, it's worth saying – AI does not have people skills. While AI can simulate empathy and emotional responses, it does not actually HAVE emotions. As such, it cannot grasp complexities, such as cultural context or personal motivations, and cannot react or advise with creativity and ethical considerations. So, it may be surprising to find out that AI has been shown to be useful when training people in durable/soft skills. IBM and Microsoft[18] have both implemented AI-driven bots to provide employees training and feedback on their soft skills. McDonald's has launched a learning portal,[19] driven by AI, which resulted in "higher employee retention, customer satisfaction, and operational efficiency." This seems counterintuitive – how can something that has no soft skills train you in soft skills? By leveraging the adaptability provided by AI, programs can create on-the-fly scenarios that move beyond basic, scripted learning and give a more personalized experience that targets an individual's needs. In addition, practicing soft skills with an AI can be more accessible (a.k.a. less performance anxiety–inducing), so individuals may be more open to taking the training in the first place.

---

[17] Morrison-Smith, S., Ruiz, J. Challenges and barriers in virtual teams: a literature review. *SN Appl. Sci.* **2**, 1096 (2020). https://doi.org/10.1007/s42452-020-2801-5

[18] https://360learning.com/blog/soft-skills-employee-training/

[19] Nadeem, M. The Golden Key: Unlocking Sustainable Artificial Intelligence Through the Power of Soft Skills! Journal of Management and Sustainability; Vol. 14, No. 2; 2024.

CHAPTER 13   THE AI ELEPHANT IN THE ROOM

Utensils Not Users

AI is amazing! The things that can be created using a generative AI like ChatGPT seem truly miraculous. The humanlike responses given by the systems can easily lead you to feel like it is aware, conjuring romantic notions of Number 5 from the movie *Short Circuit* or Robbie in the book *I, Robot* by Isaac Asimov or fears of an AI take over from the movies *Terminator* and *The Matrix*. However, the truth is that AI is a tool, not a person, and no matter how humanlike the response might seem, it is still a reflection of the person asking the question and the data set that the AI has been trained on. It lacks autonomy, and as such, it cannot come up with objectives on its own, any more than a knife in the kitchen could decide what to make for dinner.

To effectively use this tool, we have to keep in mind both its capabilities and its limitations:

- **AI Needs to Be Honed**: AI needs to be trained on a data set relevant to its ultimate task in order to be effective. As we mentioned earlier, a poorly trained AI can have biased outputs, like using a dull knife is more likely to cause injury than a sharp one.

- **AI Needs Oversight**: It's important to remember that AI needs to be checked – it requires supervision and guidance. If an AI misclassifies data or makes an incorrect recommendation, the fault lies with the developers, trainers, or users, not the AI itself.

- **Effective AI Is Specialized**: AI systems are typically designed for specific purposes, such as natural language processing, image recognition, or predictive analytics. AI is a specialized tool to solve problems, not a generalized agent capable of broad, humanlike thought.

347

CHAPTER 13   THE AI ELEPHANT IN THE ROOM

- **AI Thrives on Big Data**: AI needs Big Data as much
  as it requires electricity – that is to say that it cannot
  function without either. And, believe it or not, the
  entirety of available data (think all of the publicly
  scraped data used by AI companies) that was used
  to train models the past few years has largely been
  exhausted.[20] This fact, coupled with a number of legal
  proceedings, may determine how, how much, when,
  and where information that was previously considered
  "public" may be used to train models.

This last point about Big Data has several important ramifications. First, it means that AI cannot make sense of unique or truly rare situations that do not generate enough occurrences for it to factor into statistical models. If you think back to the opening chapter, none of the AI models that were tested were able to "imagine" something truly unique. Second, it means that there may likely come a time in the near future when advances start to plateau. Given depleted (or embargoed) sources of Big Data, some researchers have started to turn to having AI generate its own data (i.e., so-called "recursive data"), which has not yielded encouraging results.[21]

Finally, as of August 2024, the European Union Artificial Intelligence Act (EU AI Act) has gone into effect, which stipulates different levels of risk (i.e., minimal, limited, high, and unacceptable) associated with the dissemination of various AI models (and their products/output) and outlines procedures for ensuring compliance with regulations for each level and lays out sizable penalties for violations of these

---

[20] Jones, N. (2024). The AI revolution is running out of data. What can researchers do?. *Nature, 636*(8042), 290–292.

[21] Shumailov, I., Shumaylov, Z., Zhao, Y., Papernot, N., Anderson, R., & Gal, Y. (2024). AI models collapse when trained on recursively generated data. *Nature, 631*(8022), 755–759.

CHAPTER 13   THE AI ELEPHANT IN THE ROOM

policies.[22] Although this legislation is primarily aimed at developers of AI technologies, it behooves designers and anyone using AI to be aware of the broad implications of this legislation in the same way that the General Data Protection Regulation (GDPR; also from the EU) has changed how businesses think about internet data and privacy.[23]

Recap

- AI is not people. It is not creative, and it relies on the data set on which it has been trained.

- AI is not new. And neither are the concerns about its use.

- AI can process a lot of data quickly and can make inferences about that data which are useful – assuming we ask the right questions.

- AI doesn't "create" or "learn"; it "combines" and "extrapolates."

- AI has a lot of potential, but it is neither going to save us or bury us on its own – it's just a model (like Camelot in Monty Python's Holy Grail).

---

[22] The European Parliament and The Council of the European Union. 2024. "Regulation (EU) 2024/1689 of the European Parliament and of the Council of 13 June 2024 laying down harmonised rules on artificial intelligence and amending Regulations (EC) No 300/2008, (EU) No 167/2013, (EU) No 168/2013, (EU) 2018/858, (EU) 2018/1139 and (EU) 2019/2144 and Directives 2014/90/EU, (EU) 2016/797 and (EU) 2020/1828 (Artificial Intelligence Act) (Text with EEA relevance)." *Official Journal of the European Union.*
[23] https://gdpr-info.eu/

CHAPTER 13   THE AI ELEPHANT IN THE ROOM

- Humans need to make the effort to understand context to avoid making the same mistakes as AI.

- Because humans designed the vast library of information that AI uses for results, you need to know that it may (and likely will) introduce these biases into design.

- Soft skills (or durable skills) become more important as AI becomes more prevalent. While these are areas that AI can help train, they are not areas that AI can fill.

Before You Go...

So many potential stories to choose from for AI! As generative AI and smart systems are still being refined, we are treated to any number of mishaps from the disgusting but amusing (the family robot vacuum spreading dog feces all over the house) to the tragic (a young person taking the advice of AI and committing suicide). Understanding its limitations is key – it does not have context or inference, merely extrapolation. Hence, when a Google user typed "How many rocks shall I eat," the AI summary provided "According to geologists at UC Berkley, you should eat at least one small rock a day" – a result it pulled from the Onion, a well-known satirical news outlet.

As we've stressed repeatedly in this book, AI can be a great help to us – but it is just that, a helper. There is really no substitute for the wonderfully weird human psyche and all of the foibles therein. Understanding people – what we can and can't (I'm looking at you multitasking!) do – is the key to designing better products, processes, and interactions, and to helping us write and create art (while we let AI do the laundry).

Index

A

Artificial intelligence (AI), 1, 31
  dark patterns, 76, 77
  definition, 33
  design decisions, 1
  empathy/emotional
     responses, 346
  "garbage in, garbage out",
     339, 340
  human experience, 116–118
  image generators, 2, 3
  LLMs (*see* Large language
     models (LLMs))
  machine learning, 33, 34
  management issue, 344
  memory process, 213
  Moore's Law, 337–339
  neural networks, 34, 35
  performance reviews, 32, 33
  personification, 340–342
  predictive policing, 38
  psychological researchers, 340
  research (*see* Research process)
  science fiction, 341
  SkyNet, 38
  soft skills, 343–346
  terminal condition, 39
  trust design, 96–102
  user manual, 1, 2
  utensils, 347–349
  virtual communication, 345
Associative learning, 282–285
Attention (Lack Thereof)
  adaptation, 134
  bottleneck/pie
    capacity models, 129
    conversations, 128
    ethernet, 126
    informational stream, 127
    models, 130
    prevailing models, 126
    prominent model, 129
    sensory information, 127
    social and contextual
     questions, 129
  cognitive processes, 131, 132
  conversational
    groupings, 133
  email messages, 134
  habit loops
    automatic process, 140
    endowment effect, 146
    external triggers, 143
    hook model, 142, 143

INDEX

Attention (Lack Thereof) (*cont.*)
    investment, 145, 146
    parts, 141
    random rewards, 144
    variable
      reinforcements, 143–145
  habits
    definition, 135
    key difference, 136
  information processing, 132
  multi-tasking, 146
    cognitive constraints, 150
    context switching, 152
    design elements, 153–155
    digital natives, 149–151
    facts, 148
    flow, 155, 156
    formulae/coding
      schemes, 155
    Gen Z, 146–149
    inattention/
      distractibility, 147
    laboratory studies, 149
    peer review/standards, 147
    time-sensitive, 153
  override distractions, 134
  overwhelming, 131
  routines, 137–140
    automaticity, 136
    benefits, 137
    definition, 135
    designs, 139
    identification, 137
    journey map, 138, 139

  reservation system, 139
  sensory inputs, 131

B

Belongingness
  approbation, 86
  categories, 82
  communal relationships, 86
  formal and informal groups, 82
  gamification (*see* Gamification
    techniques)
  group membership, 86
  human interaction, 80
  identification, 81
  interactions, 80
  long-term disposition, 85
  minor sociability, 86
  mirror neurons, 83, 84
  powerful tool
    collaborative, 91
    communication, 88–90
    interactions, 89
    multiple levels, 87
    role-playing games/
      scenarios, 90
    shared experiences, 87, 88
    team-based games, 90, 91
  psychology, 84–87
  social structures, 79
  storytelling, 84
Brain science, 19
  cognitive scientists/
    neuroscientists, 21

INDEX

collaboration, 23
components, 20
duplication, 23
engine of thought, 20–24
fluctuations, 21
frontal lobe, 21
hemispheres, 24, 25
Information Theory, 27–32
memory, 25
occipital lobe, 22, 23
parietal lobe, 22
sensations, 23
subcortical structure, 25, 26
temporal lobe, 22

C

Cognitive biases
attentional bias, 51
emotional centers, 50
endowment effect, 52
halo effect, 52, 53
heuristics, 50
objective decisions, 50
observer-expectancy effect, 53
primacy effect, 54
recency, 55
Cognitive cartography
activities, 228
availability, 234, 235
biases, 233
concepts, 228
consistency, 234
contents, 230

definition, 227
geographic markers, 229
hindsight, 234
hippocampus, 227, 228
mimic tangibility, 232, 233
mnemonic techniques, 230
recognition, 236–238
Rosy retrospection, 233
storytelling vehicles, 230
structuring flows, 231, 232
virtual objects, 232
Communication
languages (*see* Languages)
storytelling
definition, 202
emotions/visualizations, 206, 207
enhance engagement, 205
interpretation, 202
narratives/storylines, 203
perspective, 202
reservation software, 204
storage, 205
stories, 203
Cultural considerations
Ban Chang Kuih, 172
collectivistic cultures, 177
differences, 169
features, 175
Hofstede's cultural dimensions, 177
holistic processing, 175
individualism–collectivism, 170, 172

INDEX

Cultural considerations (*cont.*)
  individualistic group, 175
  indulgence, 171
  long-term orientation, 171
  marketing strategies, 176
  masculinity–femininity, 170
  McDonald's designs, 173–175
  power distance, 170
  purchasing decisions, 179
  resemblance group, 176
  rule-based processing, 175
  uncertainty avoidance (UA),
    170, 178, 179

D

Dark side patterns
  artificial intelligence, 75, 76
  bait and switch patterns, 66–68
  basket sneaking, 60
  confirm shaming, 60, 61
  disguised advertisements, 69, 70
  drip parking, 67, 68
  false urgency, 59
  forced action, 61, 62
  interactions, 56
  interface interference, 64, 65
  loss aversion bias, 67
  malware, 74, 75
  manipulation, 58
  messages/designs, 58, 59
  nagging, 70–72
  present/immediacy bias, 67
  SAAS billing, 74

sign-up process, 56
social desirability, 61
subscription, 56
subscription traps, 63, 64
trick questions, 73, 74
Data-driven system
  air temperature/shark
    attacks, 305
  empathize, 311
  empathy workshop, 310–311
  home messages, 305
  logical decisions, 303
  problematic data, 308
  respective events, 308
  statistical relationship, 303
  trends, 306–309
Decision making
  benefits, 259
  categories, 259, 260
  certainty *vs.*
    uncertainty, 261–265
  considerations, 260
  consumer preference, 267
  deferral process, 267
  designing products, 258
  direct/indirect, 259
  ecommerce, 265
  foraging information, 268, 269
  initial/ongoing investment, 258
  intangible costs, 259
  investment options/
    opportunities, 258
  market research/data
    analytics, 260

354

INDEX

neuroscientists, 264
numbers (*see* Numbers Stymie)
opportunity, 259
persuasive techniques, 270
   door in the face technique,
     271, 272
   foot-in-the-door technique,
     270, 271
   low ball technique, 273–275
   phenomenon exploits, 274
   selling cupcakes/cookies,
     274, 275
prospect theory, 261
research participants, 263
scarcity/abundance, 265–268
temporal constraints, 265
utility theory, 261
variability, 267
Design system
   adoption, 201
   architectural components, 198
   components, 198
   considerations
     capacities, 162
     cognitive impacts, 162, 163
     demographic variables, 162
     drawbacks, 164
     generational cohorts,
       163, 164
     geological terms, 162
   culture (*see* Cultural
     considerations)
   domain expertise, 165
     Clippit, 169

Clippy, 168
Goldilocks level, 167
prescriptions, 165, 166
real-time feedback, 167
scaffolding, 167
time-consuming
   method, 167
effective, 199–201
grammar rules and
   vocabulary, 198
interaction patterns, 198
large-scale systems, 199
pattern language, 198
personas *vs.*
   personae, 180–185
Design thinking
   conceptual level, 323
   data-driven (*see* Data-
     driven system)
   different stages, 327–331
   drug testing, 323
   framework, 302
   graphics-intensive views, 334
   high-end design, 333
   problems/solutions
     defining problems, 312
     differences, 314
     Google glass, 312
     Microsoft Clippy, 313
     measuring success, 319
     problem definition,
       311, 314–318
     reduction, 320
     Reebok Air Pump, 313

355

INDEX

Design thinking (*cont.*)
 solutioning relies, 321, 322
 success criteria, 318–320
 Swiffer WetJet, 313
product cycle, 327
prototype, 322
research, 331, 332
stages, 302
techniques/goals, 328
testing process, 323–329
time segments, 326
user context, 326

E

Emotions/feeling
 dark patterns, 48
 design elements, 49
 disadvantages, 46
 factors, 46
 selective pressures, 47
 social structures, 47
 See also Feeling/emotions

F

Feeling/emotions
 anticipation, 44
 cognitive biases, 50–55
 external/internal context, 44
 oatmeal raisin cookie, 44
 prosocial behaviors, 45
Forced action patterns, 62, 63

G

Gamification techniques
 badges/medals, 94–96
 design process, 92
 elements/principles, 92
 game-design techniques, 92
 leaderboards, 96
 points and scoring systems, 94
 scenarios, 92, 93
 workflow/scenario, 93
Generalization, 11, 25, 213

H

Human experience
 adaptations, 110
 design perspective, 111, 112
 experience, 107–109
 internal clocks, 112
 limitations, 105
 limits in and out
  curb cuts, 122
  limitations, 119
  rage clicking, 119–121
  typewriters, 122
  universal usability, 121
  working process, 118
 perception/context, 109
 physical senses, 109
 ramifications, 114
 sensation/perception, 108–110
 sensory information, 106
 temporal perception, 115

INDEX

textural variations, 110
time/perception, 111–115
verbs, 112
vision/decent hearing, 106
vision/hearing
impairments, 116–118
Human intelligence, *see* Original
intelligence (OI)

I, J, K

Information Theory
bipolar cells, 28
chemicals, 27–29
communication systems, 27
definition, 27
hallucinations, 28
neuroplasticity, 30, 31
neurotransmitters, 29–31
plasticity, 30, 31
serotonin, 28
Instrumental conditioning
appetitive/aversive, 286
approaches, 287
associative (*see* Associative
learning)
behavioral contingency, 288
chore system, 290
conditioned behavior, 286
constraints, 293
endless variations, 292
evolutionary legacy, 293
gamification, 289
hippocampus, 292

law of effect, 286
principles, 288
reinforcement, 287, 289

L

Languages
auditory/visual symbols, 190
cognitive resources, 188
constructive process, 206
data/signals/information, 189
design system (*see*
Design system)
internalization, 190
neurological input/output, 193
perception/context/
interpretation, 197
production *vs.*
comprehension, 192–194
psychological research, 195
sensory information, 188
statistical reasoning,
191, 192
trigger warnings, 196–198
Large language models (LLMs)
adaptability, 36
ChatGPT, 37
data lakes, 35
development, 32
learning styles tests, 37
references, 36
self-assessment quizzes, 37
supervised/unsupervised
forms, 36

INDEX

Learning process
    associative learning, 282–285
    contribution, 294
    design principles, 280
    formal instruction, 198, 299
    instrumental
        conditioning, 285–290
    interleaving, 296
    massed/spaced learning, 295
    meta-analysis, 285
    neural pathways, 296
    neurological level, 295
    nonsense syllables, 294
    persuasive techniques, 280
    productive failure, 298, 299
    self-directed learning, 281
    spaced repetition learning, 295
    temporal/categoric
        domains, 296
    written information, 294
LLMs, *see* Large language
    models (LLMs)

M

Machine learning (ML), 27,
    33–34, 40
Mathematical Universe Hypothesis
    (MUH), 244
McDonald's designs, 173–175
Memory process
    cognitive cartography, 227–238
    definition, 210, 211
    different expiration

    categories, 216
    chemical reactions, 217
    definition, 216
    duplication
      (reduplication), 217
    echoic and iconic
      memories, 217
    limitation, 218
    neural signals, 217
    physical process, 218
    short-term memory,
      218–220
    temporal categories, 216
  generalizations, 213
  neurotransmission (*see*
    Neurotransmission)
  personal experience, 210
  procedural/semantic
    memory, 212
  reconstruction, 213
  semantic memory, 214–216
ML, *see* Machine learning (ML)
Moore's Law, 337–339
MUH, *see* Mathematical Universe
    Hypothesis (MUH)

N

Neural networks, 34–35, 192
Neuroscientists, *see* Brain science
Neurotransmission
    anterograde amnesia, 224
    chunking, 223
    images, 220

long-term memory, 220
memorable designs
  chunk information, 226
  cognitive resources, 225
  familiarity, 224
  philosophical/existential
    ramifications, 224
  moment-to-moment
    experiences, 223
  phonological loop, 221, 222
  random series, 222
  real-life practicalities, 222
  short-term memory, 220
  visuospatial sketchpad, 221
Neurotransmitters, 27–29, 44, 217
Numbers Stymie
  accountants, 254, 255
  algebra/calculus/
    statistics, 244
  conjunctive fallacy, 253
  Danny, 251
  data/design, 255–257
  decisions, 244
  descriptions, 251–253
  feelings, 251
  magnitude system, 245
  more *vs.* less, 244, 245
  naturalistic observation, 246
  numeric/probabilistic
    patterns, 248–250
  objective probabilities, 250
  smart marketers, 245
  stereotypes, 254
  subjective probabilities, 250

universal language, 244
winning ticket, 247

O

Original intelligence (OI), 1, 17, 32,
  39, 116, 339

P, Q

Pentland, Alex, 99
Personas *vs.* personae
  abstraction, 181
  archetype, 181
  characteristics, 183
  deductive methods, 183
  experience/feedback, 181
  inductive methods, 182, 183
  product development, 180
  quantitative data, 183
  representation, 180
  segment, 180

R

Research process
  antiformula arguments, 13
  building blocks, 14
  business model, 5
  coupling conversations, 4
  deductive methods, 7
  design process, 12–15
  findings, 16–18
  foundational studies, 14

INDEX

Research process (*cont.*)
    generalizability, 6
    guerrilla testing, 4
    hypothesis testing, 7
    inductive approaches, 6
    intuitive, 7
    misconception, 4
    monetary compensation, 5
    operationalize, 7–10
    particular role/context, 10
    peer review process, 5
    personal experience, 3
    physical checklists, 9
    replication, 17
    scientist, 8–11
    sociopathic tendencies, 12
    user studies, 10–12

S

Scoring system, 94

Self-correcting, 338–339
Sensory information, 26, 105, 106,
    127, 131, 188
Shannon, Claude, 27
Software as a Service (SAAS), 74
Spaced repetition
    learning (SRL), 295
Sperry, Richard, 24

T, U, V, W, X, Y, Z

Thorndike, Edward, 53, 285, 286
Trust design
    cults vs. communities, 99–102
    definition, 99
    echo chamber, 100
    effective community, 99, 100
    hallucinations, 96
    overidentification, 100
    participants, 100, 102
    principles, 98

Printed in the United States
by Baker & Taylor Publisher Services